HARNESSING THE INVISIBLE

The Science and Applications of Rays from Alpha to X

HARIKUMAR V T

PREFACE

In a world where science and technology increasingly shape our everyday experiences, much of what drives our progress remains unseen, yet profoundly impactful. The invisible rays that surround us, from alpha and beta particles to X-rays and gamma rays, play an essential role in advancing our understanding of the universe, improving healthcare, revolutionizing communication, and protecting our safety. Yet, the science behind these rays and their applications is often taken for granted, hidden in the shadows of our technological landscape.

"Harnessing the Invisible: The Science and Applications of Rays from Alpha to X" aims to illuminate the world of these invisible rays and make their significance accessible to all. This book explores the origins, production, and practical uses of a wide array of electromagnetic and particle rays, revealing how these forces, which remain unseen to the naked eye, are harnessed to benefit society. Our journey through this invisible realm uncovers fascinating scientific breakthroughs, cutting-edge technological innovations, and the pivotal role these rays play in shaping our future.

The idea for this book was inspired by the rapid advancements we've witnessed in fields like medicine, where X-rays provide life-saving diagnostics, and cancer treatments leverage gamma rays with pinpoint precision. Meanwhile, the use of radio waves in wireless communication connects billions of people across the globe, and infrared technology transforms how we view and monitor our environment. Each of these rays has a story, and each one represents a triumph of human ingenuity and scientific curiosity.

This book is organized to offer both a comprehensive understanding and an engaging exploration of each type of ray. We delve into their historical discoveries, the underlying physics that govern their behavior, and their myriad applications. Readers will also gain insights into the ongoing research and future potential of these rays, which continue to hold promise for breakthroughs in fields like quantum computing, space exploration, and environmental monitoring.

"Harnessing the Invisible" is written for anyone with a curiosity about the unseen forces that shape our world. Whether you are a science enthusiast, a student, a professional in the field, or simply someone intrigued by the mysteries of the natural world, this book will open your eyes to the vast potential of these invisible rays. By understanding how we have learned to produce and utilize these rays, we can better appreciate the boundless possibilities they offer for a brighter and more connected future.

Join me as we embark on this journey to explore the science and transformative power of the rays that, although invisible, illuminate our path to progress.

COPYRIGHT WARNING

Copyright © 2024 HARIKUMAR V T All rights reserved.

No part of this publication may be reproduced, distributed, or transmitted in any form or by any means, including photocopying, recording, or other electronic or mechanical methods, without the prior written permission of the publisher, except in the case of brief quotations embodied in critical reviews and certain other noncommercial uses permitted by copyright law.

The author and publisher disclaim all responsibility for any liability, loss, or risk, personal or otherwise, which may be incurred as a consequence, directly or indirectly, of the use and application of any of the contents of this book.

For permissions requests, inquiries about licensing, and other copyright-related matters, please contact:

HARIKUMAR V T

[vtharipnra@gmail.com]

Thank you for respecting the hard work and intellectual property rights of the author.

CONTENTS

1. The World beyond Sight: An Introduction to Invisible Rays

2. The Electromagnetic Spectrum Unveiled: From Radio Waves to Gamma Rays

3. Alpha Rays: The Power and Perils of Ionizing Particles

4. Beta Rays: Medical Marvels and Nuclear Mysteries

5. Gamma Rays: Harnessing High-Energy Radiation for Healing and Exploration

6. X-Rays: Revolutionizing Medical Imaging and Beyond

7. Ultraviolet Rays: The Light We Cannot See and Its Many Faces

8. Infrared Rays: From Heat Vision to Cutting-Edge Technology

9. Microwaves: Connecting the World and Exploring the Universe

10. Radio Waves: The Foundation of Global Communication

11. The Physics of Invisible Rays: Understanding Wavelengths, Frequencies, and Energy

12. Historical Discoveries: The Pioneers Who Revealed the Unseen

13. Radiation and Health: The Science of Safety and Risk Management

14. Rays in Medicine: Transforming Diagnostics and Treatments

15. Industrial Applications: From Material Testing to Quality Control

16. Rays in Scientific Research: Unveiling the Secrets of the Universe

17. Communication Breakthroughs: The Role of Radio and Microwaves in Modern Society

18. Environmental Monitoring: Using Rays to Study and Protect Our Planet

19. Security and Defense: How Rays Keep Us Safe from Hidden Threats

20. The Future of Ray Technology: Innovations on the Horizon

21. Ethical Considerations and Safety Regulations: Balancing Progress with Responsibility

1. THE WORLD BEYOND SIGHT

An Introduction to Invisible Rays

Our understanding of the universe has always been limited by the capabilities of our senses. The human eye, though incredibly complex, is able to perceive only a narrow sliver of the electromagnetic spectrum: visible light. What lies beyond our visual range is a world teeming with energy, from low-frequency radio waves that power our communications to high-energy gamma rays produced by cosmic events. This invisible world has fascinated scientists for centuries and continues to inspire technological advancements and groundbreaking discoveries.

In this chapter, we embark on a journey to explore the realm of invisible rays, understanding what they are, how they were discovered, and the remarkable ways they impact our daily lives. From the practical to the profound, the science of invisible rays influences everything from the gadgets we use to our understanding of the cosmos.

The Invisible Spectrum: A Hidden Reality

The electromagnetic spectrum encompasses all forms of electromagnetic radiation, ranging from long-wavelength radio waves to short-wavelength gamma rays. In between lies a variety of rays, including microwaves, infrared, ultraviolet, X-rays, and gamma rays, each with unique properties and applications. The term "invisible rays" refers to any type of electromagnetic radiation that the human eye cannot detect, as well as particle rays such as alpha, beta, and gamma rays, which belong to the domain of nuclear physics.

Our visible spectrum is confined to wavelengths between approximately 400 and 700 nanometers, encompassing the colors red through violet. Yet, outside these wavelengths lies a wealth of information and potential. The invisible rays are critical to a host of technological and scientific endeavors, allowing us to see the world in ways that would otherwise be impossible.

The Discovery of Invisible Rays

The quest to understand invisible rays began with the gradual realization that light, as perceived by the human eye, was just one small part of a much larger spectrum. Early breakthroughs came in the 19th century, with the discovery of infrared and ultraviolet light.

Infrared Rays: The Heat We Can't See

In 1800, astronomer William Herschel discovered infrared rays while experimenting with sunlight and a prism. By using thermometers to measure the temperature of the different colors of light, Herschel found that temperatures increased even beyond the red part of the visible spectrum, in an area where no visible light was present. This marked the discovery of infrared radiation, an invisible ray associated with heat. Today, infrared technology is used in everything from thermal imaging and night-vision devices to remote controls and climate studies.

Ultraviolet Rays: Beyond the Violet

The following year, in 1801, Johann Wilhelm Ritter discovered ultraviolet light. Ritter observed that invisible rays just beyond the violet part of the spectrum were capable of darkening silver chloride, a chemical used in early photographic processes. These rays were named "ultraviolet" because of their position beyond the violet end of the visible spectrum. Ultraviolet rays play a crucial role in medicine, sanitation, and the study of stars, but they are also a double-edged sword, as overexposure to UV radiation can cause skin damage and cancer.

X-Rays: A Medical Revolution

Fast forward to 1895, when German physicist Wilhelm Conrad Roentgen discovered X-rays. While experimenting with cathode rays, Roentgen noticed that a fluorescent screen glowed even when shielded from visible light. He deduced that an unknown type of ray was passing through the materials and named it "X-ray," with "X" representing the unknown. X-rays revolutionized medicine by providing a non-invasive way to see inside the human body. Today, they are an indispensable tool in diagnosing broken bones, detecting diseases, and even aiding in complex surgeries.

Radio Waves: The Birth of Wireless Communication

The discovery of radio waves was another milestone that transformed human communication. In the late 19th century, James Clerk Maxwell predicted the existence of electromagnetic waves, and Heinrich Hertz later confirmed their presence. Guglielmo Marconi harnessed these waves to create the first long-distance wireless communication system, laying the foundation for modern radio, television, and wireless networks.

The Science of Electromagnetic Waves

To appreciate the diversity of invisible rays, it is essential to understand the basic properties of electromagnetic waves. All electromagnetic waves travel at the speed of light in a vacuum but differ in wavelength and frequency. The longer the wavelength, the lower the frequency, and vice versa. This relationship governs the energy carried by the wave, with higher frequencies corresponding to greater energy levels.

The Electromagnetic Spectrum

The electromagnetic spectrum is divided into several regions, each with unique properties:

Radio Waves: With the longest wavelengths and lowest frequencies, radio waves are used for communication, including radio broadcasts, television, and mobile phones.

Microwaves: Shorter than radio waves, microwaves are used in cooking, radar, and satellite communication.

Infrared Rays: Known for their association with heat, infrared rays have numerous applications, from thermal imaging to remote controls.

Visible Light: The narrow band of the spectrum that our eyes can perceive.

Ultraviolet Rays: These rays can be beneficial, as they help produce vitamin D, but they can also be harmful, causing sunburn and increasing cancer risk.

X-Rays: With higher energy levels, X-rays penetrate soft tissues and are invaluable in medical imaging and material analysis.

Gamma Rays: The most energetic rays in the spectrum, gamma rays are produced by nuclear reactions and are used in cancer treatment and astrophysical research.

Particle Rays: Alpha, Beta, and Gamma Rays

While the electromagnetic spectrum represents one facet of the invisible world, particle rays like alpha, beta, and gamma rays fall into the domain of nuclear physics. These rays are emitted during radioactive decay and are essential to understanding atomic processes.

Alpha Rays: Comprising two protons and two neutrons, alpha particles are relatively heavy and do not penetrate far into materials. They are dangerous if ingested or inhaled but are otherwise easily shielded by a piece of paper or the skin.

Beta Rays: Consisting of high-energy electrons or positrons, beta rays have greater penetration power than alpha particles and are used in medical treatments and radiation therapy.

Gamma Rays: These are high-energy photons emitted from nuclear decay, with significant penetration power. Gamma rays are used in medical treatments and are also a focus of astrophysical research to study cosmic phenomena.

Applications of Invisible Rays

Invisible rays are integral to countless technologies and scientific advancements. Here are some of the most noteworthy applications:

Medical Imaging and Treatment

X-rays and gamma rays have revolutionized medical science. X-ray machines provide critical diagnostic images, while gamma rays are used in cancer treatments like radiotherapy. Additionally, positron emission tomography (PET) scans use beta rays to visualize metabolic processes in the body, aiding in disease diagnosis and treatment planning.

Infrared technology is used for thermal imaging, which helps detect fevers and monitor blood flow. Ultraviolet rays, although dangerous in excess, are used to sterilize medical equipment and treat certain skin conditions.

Communication and Navigation

Radio waves and microwaves are the backbone of modern communication systems, enabling everything from radio and television broadcasts to cellular networks and satellite communication. GPS technology relies on microwaves to provide accurate location data, transforming navigation for people and industries worldwide.

Environmental Monitoring

Infrared and ultraviolet rays are used to monitor the environment and study climate change. Satellites equipped with infrared sensors measure heat emissions from the Earth's surface, while ultraviolet detectors track ozone layer depletion. Understanding how invisible rays interact with the atmosphere is critical for developing strategies to combat environmental challenges.

Industrial and Security Applications

Invisible rays are also essential in industry and security. Gamma rays and X-rays are used in non-destructive testing to inspect the integrity of materials, such as bridges, buildings, and pipelines. Airport security scanners use X-rays to detect prohibited items in luggage. Ultraviolet light is employed to detect counterfeit currency and sanitize surfaces in public spaces.

Space Exploration

In space exploration, gamma rays and X-rays are used to study celestial bodies and phenomena. Gamma-ray bursts, for example, are the most powerful explosions in the universe, providing insights into the formation of black holes and the behavior of matter under extreme conditions. Radio waves help astronomers study distant galaxies and map cosmic background radiation, revealing the universe's history and structure.

The Future of Invisible Rays

Research into invisible rays continues to yield new discoveries and applications. Innovations such as terahertz imaging, which lies between microwaves and infrared on the spectrum, promise to revolutionize medical diagnostics and security screening. Quantum communication, leveraging the unique properties of photons, could redefine secure data transmission.

The world beyond sight is teeming with rays that have transformed our understanding of the universe and fueled technological advancements across various fields. From radio waves that keep us connected to gamma rays that reveal the secrets of the cosmos, these invisible forces are as awe-inspiring as they are practical. As we continue to explore and harness the power of invisible rays, the potential for new breakthroughs and innovations is limitless.

"Harnessing the Invisible: The Science and Applications of Rays from Alpha to X" invites you to delve deeper into this hidden world. Each chapter will reveal the incredible ways in which these rays impact our lives, from the mundane to the extraordinary, showcasing the vast potential of a world we can't see but have come to master.

This chapter provides an engaging foundation for exploring the science and applications of invisible rays in the subsequent chapters.

2. THE ELECTROMAGNETIC SPECTRUM UNVEILED

From Radio Waves to Gamma Rays

Our universe is awash in a sea of electromagnetic waves, invisible streams of energy that travel across the cosmos, connecting galaxies and powering our modern world. Yet, our human senses are only able to perceive a small fraction of these waves, in the form of visible light. The electromagnetic spectrum, which encompasses everything from long radio waves to incredibly energetic gamma rays, is fundamental to understanding both the mysteries of the universe and the technologies we rely on daily.

In this chapter, we will explore the electromagnetic spectrum in detail, unveiling the diverse forms of electromagnetic radiation, their properties, and their applications. We will examine how each part of the spectrum is unique, yet interconnected, and why this understanding is crucial to the scientific and technological advancements that shape our world.

The Nature of Electromagnetic Waves

Before diving into the specifics of the electromagnetic spectrum, it is essential to understand what electromagnetic waves are and how they are created. Electromagnetic waves are disturbances in electric and magnetic fields that propagate through space. They are generated when charged particles, such as electrons, accelerate. This acceleration creates oscillating electric and magnetic fields that move through space at the speed of light.

The key parameters that characterize electromagnetic waves are wavelength and frequency. Wavelength refers to the distance between successive peaks of the wave and is typically measured in meters or nanometers. Frequency, measured in hertz (Hz), indicates how many wave cycles pass a given point per second. These two properties are inversely related: as the wavelength decreases, the frequency increases, and vice versa. The energy of an electromagnetic wave is directly proportional to its frequency, meaning higher-frequency waves carry more energy.

The electromagnetic spectrum spans an enormous range of wavelengths and frequencies, from long radio waves with wavelengths that can be as large as kilometers to high-energy gamma rays with wavelengths smaller than an atom's nucleus. Understanding this spectrum is like having a universal translator for the language of the universe, revealing information hidden from our eyes and even expanding the boundaries of human perception.

The Regions of the Electromagnetic Spectrum

The electromagnetic spectrum can be divided into several distinct regions, each with its own set of characteristics and uses. These regions include, in order of increasing frequency and decreasing wavelength: radio waves, microwaves, infrared, visible light, ultraviolet, X-rays, and gamma rays. Let's explore each of these regions in detail.

1. Radio Waves: The Foundation of Communication

Wavelength: 1 millimeter to 100 kilometers
Frequency: 3 kHz to 300 GHz

Radio waves have the longest wavelengths and the lowest frequencies in the electromagnetic spectrum. They are ubiquitous in our daily lives, facilitating communication over vast distances. When we turn on a radio, connect to a Wi-Fi network, or use a GPS device, we are harnessing the power of radio waves.

Applications of Radio Waves:

Broadcasting: AM and FM radio stations transmit audio signals using radio waves, while television broadcasting relies on them to deliver video content.

Communication: Cell phones, walkie-talkies, and satellite communication systems use radio waves for data transmission.

Navigation: The Global Positioning System (GPS) uses radio waves to provide accurate location information, essential for navigation.

Astronomy: Radio telescopes capture radio waves emitted by celestial objects, helping astronomers study phenomena like pulsars and distant galaxies.

Radio waves are instrumental in expanding our understanding of the universe and enabling global communication networks.

2. Microwaves: From Cooking to Cosmic Exploration

Wavelength: 1 millimeter to 1 meter
Frequency: 300 MHz to 300 GHz

Microwaves are shorter than radio waves and are best known for their use in microwave ovens. However, their applications extend far beyond the kitchen. Microwaves are crucial in telecommunications and even in the study of the early universe.

Applications of Microwaves:

Cooking: Microwave ovens use microwaves to heat food. These waves excite water molecules, generating heat and cooking the food from the inside out.

Communication: Microwaves carry signals for mobile phones, Wi-Fi, and satellite communication. They are also used in radar technology, which detects and tracks objects like aircraft and weather formations.

Astronomy: The cosmic microwave background radiation, a relic from the Big Bang, provides vital clues about the origins of the universe.

Microwave technology has revolutionized communication, making it faster and more efficient, while also giving us a glimpse into the universe's infancy.

3. Infrared Radiation: The Heat of the Invisible

Wavelength: 700 nanometers to 1 millimeter
Frequency: 300 GHz to 430 THz

Infrared radiation lies just beyond the visible spectrum, and it is most commonly associated with heat. Any object that emits heat radiates infrared energy, from the warmth of a human body to the thermal glow of a distant star.

Applications of Infrared Radiation:

Thermal Imaging: Infrared cameras detect heat and are used in night-vision technology, medical diagnostics, and surveillance.

Remote Controls: Many electronic devices, such as televisions, use infrared signals to communicate wirelessly with remote controls.

Astronomy: Infrared telescopes observe celestial objects obscured by cosmic dust, revealing stars, galaxies, and nebulae invisible in other wavelengths.

Climate Science: Infrared satellites monitor the Earth's heat emissions, helping scientists study climate change and weather patterns.

Infrared technology is indispensable in fields ranging from medicine to astronomy, offering insights into both the human body and the vastness of space.

4. Visible Light: The Narrow Band We See

Wavelength: 400 to 700 nanometers
Frequency: 430 THz to 770 THz

Visible light is the only part of the electromagnetic spectrum that the human eye can detect. It consists of the colors we see in a rainbow: red, orange, yellow, green, blue, indigo, and violet. Each color corresponds to a different wavelength, with red having the longest wavelength and violet the shortest.

Applications of Visible Light:

Illumination: From sunlight to electric bulbs, visible light is essential for vision and daily activities.

Photography and Film: Cameras capture images using visible light, creating pictures and videos that document and interpret our world.

Science and Medicine: Visible light is used in optical microscopes to examine cells and tissues, and lasers play a role in surgeries and scientific research.

Despite being a small fraction of the spectrum, visible light is vital to life on Earth, as it drives photosynthesis and influences our perception of the environment.

5. Ultraviolet Rays: The Double-Edged Sword

Wavelength: 10 to 400 nanometers
Frequency: 770 THz to 30 PHz

Ultraviolet (UV) rays have shorter wavelengths than visible light and are divided into three categories: UVA, UVB, and UVC. While UV rays are necessary for vitamin D production in humans, overexposure can lead to skin damage and cancer.

Applications of Ultraviolet Rays:

Medical Sterilization: UV light is used to kill bacteria and viruses, making it essential for sanitizing medical equipment.

Forensics: UV light helps detect biological substances like blood and saliva, assisting crime scene investigations.

Astronomy: UV telescopes study young, hot stars and the behavior of galaxies.

UV rays have both beneficial and harmful effects, necessitating a balance between exposure and protection.

6. X-Rays: Peering Inside the Human Body

Wavelength: 0.01 to 10 nanometers
Frequency: 30 PHz to 30 EHz

X-rays have higher energy than ultraviolet rays and are capable of penetrating soft tissue, making them invaluable for medical imaging.

Discovered by Wilhelm Roentgen in 1895, X-rays transformed the medical field by providing a non-invasive way to see inside the human body.

Applications of X-Rays:

Medical Imaging: X-rays are used to diagnose broken bones, infections, and certain diseases. Advanced techniques, such as CT scans, create detailed cross-sectional images of the body.

Security: Airports use X-ray machines to inspect luggage and detect concealed objects.

Astronomy: X-ray telescopes study high-energy phenomena, such as black holes and supernovae.

X-rays are a cornerstone of modern medicine and have revolutionized our understanding of both the human body and the universe.

7. Gamma Rays: The Universe's Most Energetic Rays

Wavelength: Less than 0.01 nanometers
Frequency: Greater than 30 EHz

Gamma rays are the most energetic and penetrating form of electromagnetic radiation. They are produced by radioactive decay, nuclear reactions, and cosmic phenomena like supernovae. Gamma rays have both destructive and beneficial uses.

Applications of Gamma Rays:

Cancer Treatment: Gamma rays are used in radiotherapy to target and destroy cancer cells while sparing surrounding healthy tissue.

Sterilization: Gamma radiation sterilizes medical equipment and food by killing bacteria and pathogens.

Astrophysics: Gamma-ray telescopes observe the universe's most violent events, such as gamma-ray bursts, which are the most energetic explosions observed in the cosmos.

Gamma rays are both awe-inspiring and formidable, pushing the boundaries of science and technology.

The Unifying Thread: Electromagnetic Radiation in Our Lives

The electromagnetic spectrum, in all its diversity, connects every aspect of our world. From the gentle hum of radio waves to the penetrating energy of gamma rays, electromagnetic radiation powers communication, fuels scientific discovery, and even keeps us warm on a sunny day. It has revolutionized medicine, astronomy, and technology, enabling humanity to explore realms both large and small, near and far.

Yet, with great power comes great responsibility. As we harness electromagnetic waves for progress, we must also consider their effects on health and the environment. Balancing innovation with safety is key to building a sustainable future where the benefits of this invisible energy can be shared by all.

In the chapters that follow, we will explore how scientists and engineers have harnessed these waves to develop technologies that shape our world and discover new frontiers in our understanding of the universe. The electromagnetic spectrum is, indeed, a bridge between the known and the unknown—a testament to the boundless potential of human curiosity and ingenuity.

3. ALPHA RAYS

The Power and Perils of Ionizing Particles

When Henri Becquerel discovered the mysterious phenomenon of radioactivity in 1896, it marked the beginning of an era that would fundamentally change our understanding of atomic physics. Following this discovery, the early 20th century was a time of great revelations in the world of subatomic particles. Among the key players in this invisible world are alpha rays, powerful yet relatively slow-moving forms of ionizing radiation that have influenced scientific research, medical advancements, and our awareness of radiation hazards. This chapter will delve into what alpha rays are, their origins, their unique characteristics, and the profound impact they have had, both beneficial and harmful.

What Are Alpha Rays?

Alpha rays, or alpha particles, are a type of ionizing radiation emitted from the nucleus of an atom during radioactive decay. These particles consist of two protons and two neutrons, bound together to form a helium-4 nucleus. This composition makes alpha particles relatively massive compared to other forms of radiation, such as beta particles and gamma rays. Due to their relatively large mass and positive charge, alpha particles interact strongly with matter, losing energy quickly as they travel through a medium.

The charge of an alpha particle is +2, owing to its two protons, and it has a relatively high energy, typically in the range of 4 to 8 MeV (mega-electronvolts). However, because of their size and the strong interaction with surrounding atoms, alpha particles have a short range and are easily stopped by a sheet of paper, a few centimeters of air, or the outer layer of human skin. Despite their limited penetration, alpha particles are not to be underestimated; their ionizing power can have serious biological effects when the source of radiation is internalized through inhalation, ingestion, or open wounds.

The Origins of Alpha Rays

Alpha radiation is commonly emitted by heavy, unstable nuclei, such as those of uranium-238, radium-226, and polonium-210. These elements are found in nature as part of the decay chains of uranium and thorium. For instance, uranium-238 undergoes a series of radioactive transformations, one of which is the emission of an alpha particle, gradually leading to the formation of stable lead-206.

The discovery of alpha radiation is credited to Ernest Rutherford, who, in 1899, distinguished alpha rays from beta rays based on their different penetrating abilities. Rutherford's work laid the foundation for the field of nuclear physics, and his experiments with alpha particles eventually led to the famous gold foil experiment, which demonstrated that atoms consist of a dense, positively charged nucleus surrounded by empty space. This discovery revolutionized the atomic model and underscored the significance of alpha radiation in advancing scientific knowledge.

Properties of Alpha Rays

Alpha particles have a number of distinct properties that differentiate them from other types of radiation. These properties include their mass, charge, energy, ionizing power, and penetrating ability. Let's examine these characteristics in detail.

Mass and Structure: As mentioned earlier, alpha particles are relatively heavy because they consist of two protons and two neutrons. This mass gives alpha rays significant kinetic energy but also makes them less capable of penetrating materials compared to lighter particles like beta rays.

Charge: The +2 charge of alpha particles causes them to interact strongly with electrons in the atoms they encounter. This strong interaction leads to a high rate of ionization, making alpha rays exceptionally effective at ionizing matter.

Energy: Alpha particles are emitted with a specific energy that is characteristic of the radioactive isotope from which they originate. The energy levels of alpha particles are much higher than those of most beta particles, but this energy is rapidly dissipated as they travel through matter.

Ionizing Power: The ability of alpha rays to ionize atoms is one of their most significant properties. As alpha particles pass through a medium, they strip electrons from atoms, creating ions. This ionization process can disrupt chemical bonds and damage biological molecules, including DNA.

Penetrating Ability: Despite their high energy, alpha particles have very low penetration power. As a result, they are easily stopped by barriers, such as a few centimeters of air or human skin. However, if alpha-emitting materials are inhaled, ingested, or enter the body through a wound, they can deposit large amounts of energy in a small area, causing significant damage to living tissues.

Sources of Alpha Radiation

Alpha radiation is emitted by a variety of naturally occurring and man-made radioactive elements. The most common sources include:

Uranium-238: A naturally occurring radioactive element found in soil, rocks, and water. Uranium-238 undergoes alpha decay to form thorium-234.

Radium-226: Discovered by Marie and Pierre Curie, radium-226 emits alpha particles and is found in uranium ores. It was once used in luminescent paint for watch dials and instrument panels, but its use has been discontinued due to safety concerns.

Polonium-210: A highly radioactive element discovered by Marie Curie. Polonium-210 is an alpha emitter that has been used in scientific research and, infamously, as a poison in high-profile assassination cases.

Americium-241: A synthetic element used in smoke detectors. Americium-241 emits alpha particles, which ionize air molecules to help detect smoke particles.

These sources highlight the diverse nature of alpha-emitting materials and their prevalence in both natural and industrial settings.

Applications of Alpha Radiation

Despite the hazards associated with alpha radiation, it has a variety of important applications across different fields, including medicine, industry, and scientific research. Let's explore some of the ways alpha radiation is harnessed for beneficial purposes.

Medical Applications: Alpha radiation plays a crucial role in certain medical treatments. Alpha-emitting isotopes, such as radium-223 and actinium-225, are used in targeted alpha therapy (TAT) to treat specific types of cancer. This therapy involves delivering alpha-emitting isotopes directly to cancer cells, where the high ionizing power of alpha particles destroys malignant cells while minimizing damage to surrounding healthy tissue.

Smoke Detectors: One of the most common household uses of alpha radiation is in ionization smoke detectors. These devices contain a small amount of americium-241, which emits alpha particles. The particles ionize the air in a detection chamber, creating a current. When smoke enters the chamber, it disrupts the ionization process, triggering an alarm.

Space Exploration: Alpha radiation is used in radioisotope thermoelectric generators (RTGs), which power spacecraft. The heat generated by the decay of alpha-emitting isotopes is converted into electricity, providing a reliable power source for deep-space missions where solar energy is not feasible.

Material Analysis: In a technique called alpha spectrometry, alpha particles are used to analyze the composition of materials. This method is valuable in environmental science for detecting and measuring the levels of alpha-emitting isotopes in soil, water, and air samples.

These applications demonstrate the versatility of alpha radiation and its contributions to both technology and human health. However, the use of alpha radiation must be carefully managed to prevent exposure and minimize risks.

The Dangers of Alpha Radiation

While alpha particles are easily blocked by external barriers, they pose a significant health risk if alpha-emitting materials enter the body. The ionizing power of alpha particles can cause severe damage to living tissues, especially if radioactive materials are inhaled, ingested, or enter through open wounds. This damage can lead to cellular mutations, cancer, and acute radiation poisoning.

Health Effects: The biological damage caused by alpha radiation is due to its ability to ionize atoms within cells. This ionization can break chemical bonds, disrupt DNA, and interfere with normal cellular functions. Prolonged or intense exposure can lead to radiation sickness, an increased risk of cancer, and damage to organs.

Historical Incidents: There have been several high-profile cases involving the dangers of alpha radiation. The most notable is the Radium Girls tragedy in the early 20th century. Women who painted watch dials with radium-based luminescent paint were exposed to alpha radiation when they inadvertently ingested radium, leading to severe health problems and death. Another case is the assassination of Alexander Litvinenko in 2006, who was poisoned with polonium-210, an alpha-emitting isotope.

Regulations and Safety Measures: To protect against the hazards of alpha radiation, strict regulations govern the handling, storage, and disposal of alpha-emitting materials. Safety measures include wearing protective gear, using ventilation systems to prevent inhalation, and following protocols to minimize contamination and exposure.

The Future of Alpha Radiation Research

Ongoing research aims to harness the power of alpha radiation for new and innovative applications while minimizing the associated risks. In the field of medicine, advancements in targeted alpha therapy offer hope for more effective cancer treatments with fewer side effects. Researchers are also exploring the use of alpha particles in advanced imaging techniques and as a tool for studying the properties of matter at the atomic level.

Additionally, scientists are investigating ways to mitigate the environmental impact of alpha-emitting materials, particularly in areas affected by mining and nuclear waste. The goal is to develop technologies that can safely contain or neutralize radioactive contaminants, ensuring the protection of ecosystems and human populations.

The Dual Nature of Alpha Rays

Alpha rays epitomize the dual nature of ionizing radiation: a force that holds the potential for both destruction and healing. From their discovery at the dawn of the atomic age to their modern applications in medicine and technology, alpha particles continue to shape our understanding of the universe and our approach to harnessing the power of the atom. As we move forward, the challenge lies in balancing the benefits of alpha radiation with the imperative to protect human health and the environment. By learning from the past and investing in safe, responsible research, we can unlock the full potential of alpha rays for the betterment of society.

This exploration of alpha rays serves as a foundation for understanding the complexities of ionizing radiation and its impact on science, technology, and everyday life. As we continue our journey through the electromagnetic spectrum, we will encounter other forms of radiation, each with its own unique properties and applications.

4. BETA RAYS

Medical Marvels and Nuclear Mysteries

In the realm of ionizing radiation, beta rays stand out for their dualistic nature: powerful enough to alter human health outcomes positively through medical applications, yet mysterious and potentially hazardous when unleashed in uncontrolled environments. Understanding beta rays is essential not only for scientists and medical professionals but also for those keen to grasp how these invisible particles shape our everyday lives and the world around us.

Beta rays have played pivotal roles in both nuclear physics and medical advancements, but they also present intriguing complexities that require ongoing study and careful management. This chapter will explore the nature of beta rays, their scientific discovery, their fascinating properties, and their wide-ranging applications and risks.

What Are Beta Rays?

Beta rays are streams of high-speed electrons or positrons emitted during the radioactive decay of certain unstable atomic nuclei. These emissions fall into two categories: beta-minus (β^-) rays and beta-plus (β^+) rays. Beta-minus rays consist of negatively charged electrons, while beta-plus rays are composed of positively charged positrons, the antimatter counterparts of electrons. Both types of beta rays are forms of ionizing radiation, capable of stripping electrons from atoms and molecules and thus altering the chemical and physical structure of matter.

The emission of beta rays occurs when a neutron within an atomic nucleus transforms into a proton (beta-minus decay) or when a proton transforms into a neutron (beta-plus decay). These transformations are governed by the weak nuclear force, one of the four fundamental forces of nature. The release of beta particles during these decay processes contributes to the stability of an atom, helping it reach a more balanced state.

Beta radiation was first distinguished from other types of radiation by the early pioneers of radioactivity research. Ernest Rutherford and Marie Curie made significant contributions to the field, but it was Henri Becquerel who initially discovered radioactivity, laying the groundwork for future exploration into these subatomic phenomena.

Properties of Beta Rays

Beta rays have several unique properties that differentiate them from alpha and gamma radiation. These properties include their mass, charge, energy range, ionizing power, and penetration capability. Let's delve into these characteristics to understand how beta rays interact with the world.

Mass and Charge: Unlike alpha particles, which are relatively massive, beta particles are much lighter. Electrons have a negligible mass compared to protons and neutrons, making beta rays far more agile and less prone to interaction with matter. Beta-minus particles carry a negative charge, while beta-plus particles have a positive charge.

Energy Range: Beta particles can be emitted with a wide range of energies, from a few keV (kilo-electronvolts) to several MeV (mega-electronvolts). The energy of beta rays depends on the specific radionuclide undergoing decay. This energy range affects how deeply beta rays can penetrate materials and their effectiveness in various applications.

Ionizing Power: Beta rays have a moderate ionizing power, higher than that of gamma rays but lower than alpha particles. They can ionize atoms and molecules, potentially causing damage to living tissue. The extent of this ionization depends on the energy of the beta particles and the nature of the material they encounter.

Penetration Capability: Beta rays have a greater penetrating ability than alpha particles but are still relatively easy to shield. A few millimeters of plastic, glass, or aluminum can block beta radiation. However, beta particles can penetrate the outer layer of human skin, posing a risk if a person is exposed to beta-emitting materials.

Sources of Beta Radiation

Beta radiation is emitted by a variety of radioactive isotopes, both natural and man-made. Some common sources include:

Carbon-14: A naturally occurring isotope used in radiocarbon dating to determine the age of archaeological and geological samples. Carbon-14 undergoes beta-minus decay, emitting beta rays as it transforms into nitrogen-14.

Strontium-90: A byproduct of nuclear fission in reactors and weapons. Strontium-90 is a significant environmental contaminant and emits beta-minus rays as it decays into yttrium-90.

Iodine-131: Used in medical treatments for thyroid conditions, iodine-131 emits beta rays and is an essential tool in both diagnostics and therapy.

Fluorine-18: A beta-plus emitter used in positron emission tomography (PET) scans. Fluorine-18 decays by emitting positrons, which annihilate with electrons to produce gamma rays that can be detected to create detailed images of internal organs.

These sources illustrate the diverse roles of beta rays, from aiding scientists in uncovering the past to diagnosing and treating modern-day diseases.

Applications of Beta Rays

Beta rays have a wide range of applications, many of which have revolutionized medicine, industry, and scientific research. Let's examine how beta radiation is utilized in these fields.

1. Medical Applications

Beta radiation has become a cornerstone of nuclear medicine, where it is used for both diagnostic and therapeutic purposes.

Radiotherapy: Beta rays are used in the treatment of various cancers. Strontium-89 and yttrium-90 are beta-emitting isotopes commonly employed in radiotherapy to target and destroy malignant cells. By focusing beta radiation on a tumor, doctors can minimize damage to surrounding healthy tissue while effectively killing cancerous cells. This approach has been successful in treating bone metastases and liver cancer.

Radioisotope Therapy: Iodine-131 is a beta-emitting isotope used to treat thyroid cancer and hyperthyroidism. When administered in a controlled dose, iodine-131 accumulates in the thyroid gland, where it delivers targeted beta radiation to destroy overactive or cancerous cells. This treatment is highly effective and has been in use for decades.

Positron Emission Tomography (PET): Beta-plus emitters like fluorine-18 are essential for PET scans. When fluorine-18 decays, it emits positrons that collide with electrons in the body, producing gamma rays. These rays are detected to create detailed images of organs and tissues, helping doctors diagnose conditions and plan treatments.

2. Industrial Applications

In industry, beta rays are used for various non-destructive testing and measurement techniques.

Thickness Gauging: Beta radiation is used to measure the thickness of materials, such as paper, plastic, and metal sheets. By passing beta rays through the material and measuring the intensity of the radiation that emerges on the other side, manufacturers can ensure uniform thickness and quality control in their products.

Radiography: Beta rays are used in some forms of industrial radiography to inspect the integrity of materials and structures. This application is crucial for detecting flaws in welds, pipelines, and critical components in construction and manufacturing.

Static Elimination: Beta-emitting devices are used to eliminate static electricity in industrial processes. By ionizing the air, these devices can neutralize static charges that might otherwise cause issues in manufacturing environments.

3. Scientific Research

Beta rays have played a critical role in advancing scientific understanding, particularly in the fields of nuclear physics and cosmology.

Radiocarbon Dating: Carbon-14, a beta-emitting isotope, is used in radiocarbon dating to determine the age of organic materials. This technique has been instrumental in archaeology, paleontology, and climate science, helping researchers piece together the history of our planet and its inhabitants.

Nuclear Physics: The study of beta decay has provided valuable insights into the weak nuclear force and the nature of subatomic particles. The discovery of the neutrino, a nearly massless particle emitted alongside beta rays, was a direct result of research into beta decay. Neutrinos are now a key area of study in astrophysics and particle physics.

The Dangers of Beta Radiation

While beta rays have numerous beneficial applications, they also pose significant risks if not properly controlled. Exposure to beta radiation can have serious health consequences, particularly if beta-emitting materials are ingested or inhaled.

Health Risks: The ionizing power of beta rays can damage living tissue, leading to burns, radiation sickness, and an increased risk of cancer. The severity of these effects depends on the energy of the beta particles and the duration of exposure. For example, workers handling beta-emitting materials must take precautions to prevent skin exposure and avoid inhaling or ingesting radioactive particles.

Environmental Contamination: Accidents at nuclear power plants, such as the Chernobyl and Fukushima disasters, have released large quantities of beta-emitting isotopes into the environment. Strontium-90 and cesium-137 are among the most concerning contaminants, as they can persist in the environment for decades and accumulate in living organisms. These incidents highlight the need for stringent safety measures and effective waste management practices.

Radiation Protection: Shielding from beta radiation is relatively straightforward compared to alpha or gamma radiation. Thin layers of plastic, glass, or aluminum can effectively block beta rays. However, when beta-emitting materials are present in the air or water, more complex safety measures are required to prevent contamination and protect human health.

The Mystery of Antimatter: Beta-Plus Decay

One of the most fascinating aspects of beta radiation is beta-plus decay, which produces positrons—the antimatter counterparts of electrons. Positrons are identical to electrons in mass but carry a positive charge. When a positron encounters an electron, the two particles annihilate each other, releasing energy in the form of gamma rays. This process is the basis for PET scans and provides a glimpse into the mysterious world of antimatter.

The existence of antimatter raises profound questions about the nature of the universe. Why is there more matter than antimatter? What role does antimatter play in cosmology? These questions continue to puzzle scientists and fuel research into the fundamental forces that govern our universe.

Beta rays occupy a unique position in the world of ionizing radiation, balancing between scientific curiosity and practical utility. From life-saving medical treatments to crucial industrial applications, beta radiation has transformed numerous fields. Yet, the inherent dangers of beta rays require us to approach their use with respect and caution, ensuring that their benefits are harnessed safely and responsibly.

As we continue to explore the science of beta rays, we uncover new possibilities and challenges. The future holds promise for further advancements in medicine, industry, and research, driven by our evolving understanding of these powerful and enigmatic particles. In our journey to harness the invisible forces of nature, beta rays remind us of the delicate balance between discovery and responsibility.

This exploration of beta rays sets the stage for deeper inquiries into the electromagnetic spectrum and the powerful forces that shape our world. As we proceed, we will examine how other rays, each with its unique properties, contribute to science, technology, and our understanding of the universe.

5. GAMMA RAYS

Harnessing High-Energy Radiation for Healing and Exploration

Gamma rays, the most energetic form of electromagnetic radiation, are both a marvel and a mystery in the world of physics. These high-energy photons, with frequencies above 10^{19} Hz and wavelengths shorter than 10 picometers, possess the power to alter the very fabric of matter they encounter. The incredible potential of gamma rays has unlocked numerous applications, from life-saving medical therapies to groundbreaking insights into the cosmos. However, their immense energy also poses significant risks, requiring precise handling and meticulous safety protocols.

This chapter will delve into the nature of gamma rays, their discovery, their unique properties, and the various ways they are harnessed for human benefit and scientific advancement. We will also discuss the inherent dangers of gamma rays and the measures taken to mitigate these risks.

The Discovery of Gamma Rays

The journey to understanding gamma rays began at the dawn of the 20th century, a time of rapid advancements in nuclear physics and radiation research. The story starts with Henri Becquerel's discovery of natural radioactivity in 1896, followed closely by Marie and Pierre Curie's work on radioactive elements. In 1900, the physicist Paul Villard made a groundbreaking observation while studying radiation emitted by radium. He noted a highly penetrating form of radiation that differed from the already identified alpha and beta rays. This new form of radiation, later named gamma rays by Ernest Rutherford, would prove to be the most powerful yet enigmatic type of electromagnetic radiation.

Gamma rays arise from nuclear reactions, such as radioactive decay, nuclear fission, and nuclear fusion. They are produced when an excited atomic nucleus releases excess energy to stabilize itself. These emissions are purely electromagnetic, with no mass or charge, distinguishing gamma rays from alpha and beta particles. Despite their discovery over a century ago, gamma rays continue to be a subject of intense research, revealing ever more about the universe and offering potential for novel applications.

Properties of Gamma Rays

Gamma rays have several distinct properties that make them both powerful and dangerous. Understanding these characteristics is essential for harnessing their energy while protecting against their potentially harmful effects.

High Energy: Gamma rays are the highest-energy photons in the electromagnetic spectrum. They carry enough energy to penetrate most materials, including metals and dense matter. This energy enables gamma rays to ionize atoms and molecules, breaking chemical bonds and disrupting biological tissues.

Short Wavelengths: With wavelengths shorter than a few picometers, gamma rays can easily penetrate and pass through most materials. Their short wavelengths allow them to interact with atomic nuclei, making gamma rays useful for probing the fundamental structure of matter.

No Mass or Charge: Unlike alpha or beta particles, gamma rays are massless and carry no electric charge. This property allows them to travel long distances and penetrate deep into matter without being significantly deflected by electric or magnetic fields.

Ionizing Ability: Gamma rays are a form of ionizing radiation, meaning they can remove electrons from atoms and create ions. This ionization process can have both beneficial and harmful effects, depending on the context of exposure.

These properties make gamma rays a formidable tool for both medical and scientific purposes. Their ability to penetrate dense matter makes them invaluable in imaging and analysis, while their ionizing power is harnessed in targeted medical treatments.

Sources of Gamma Rays

Gamma rays are produced through various natural and artificial processes. Understanding these sources provides insight into the diversity of gamma-ray applications and the precautions needed to manage them safely.

Natural Sources: Gamma rays are constantly emitted from natural sources, including the radioactive decay of elements like uranium, thorium, and radium in the Earth's crust. The interaction of cosmic rays with the Earth's atmosphere also generates gamma radiation. Additionally, certain astronomical events, such as supernovae and gamma-ray bursts, are powerful natural sources of gamma rays.

Artificial Sources: Human-made sources of gamma rays include nuclear reactors, particle accelerators, and medical devices. Nuclear reactions, such as fission and fusion, release gamma radiation, making it essential to shield and contain these emissions in controlled environments. Radioactive isotopes like cobalt-60 and cesium-137 are commonly used in industrial and medical applications for their gamma-ray emissions.

Astronomical Sources: The universe is teeming with high-energy processes that produce gamma rays. Astrophysical phenomena, such as black holes, neutron stars, and supernovae, emit vast amounts of gamma radiation. The study of these cosmic gamma rays has led to remarkable discoveries about the nature of the universe, from the behavior of matter under extreme conditions to the mechanisms of star formation and galactic evolution.

Medical Applications of Gamma Rays

One of the most remarkable uses of gamma rays lies in the field of medicine, where they have revolutionized diagnostics and treatment. By precisely harnessing the power of gamma rays, doctors can both save lives and improve the quality of patient care.

1. Cancer Treatment: Gamma Knife Radiosurgery

The most notable medical application of gamma rays is in radiation therapy for cancer treatment. The Gamma Knife, a non-invasive radiosurgical device, uses focused gamma radiation to destroy cancerous tumors and abnormal tissue. By directing multiple beams of gamma rays from different angles to converge on a single target, the Gamma Knife delivers a high dose of radiation to the tumor while minimizing damage to surrounding healthy tissue. This technique is especially effective for treating brain tumors and arteriovenous malformations (AVMs).

Gamma Knife radiosurgery is a preferred option for patients with inoperable or hard-to-reach tumors. It offers several advantages, including a lower risk of infection, reduced recovery time, and a high degree of precision. The success of Gamma Knife therapy underscores the potential of gamma rays to transform medical practice, providing hope for patients with previously untreatable conditions.

2. Medical Imaging: PET Scans

Positron Emission Tomography (PET) is another critical application of gamma rays in medicine. PET scans use gamma radiation to create detailed images of the body's internal organs and tissues. The process involves the use of radioactive tracers, such as fluorodeoxyglucose (FDG), which emit positrons as they decay. When these positrons interact with electrons in the body, they produce gamma rays that are detected by the PET scanner.

PET scans are invaluable for diagnosing and monitoring various diseases, including cancer, heart disease, and neurological disorders. They provide insights into the metabolic activity of tissues, helping doctors assess the effectiveness of treatments and make informed decisions about patient care. The use of gamma rays in PET imaging has significantly improved our understanding of human physiology and disease mechanisms.

3. Sterilization of Medical Equipment

Gamma rays are also used to sterilize medical equipment and supplies. The ionizing power of gamma radiation is highly effective at killing bacteria, viruses, and other pathogens, making it an essential tool for ensuring the safety of medical devices. Unlike heat-based sterilization methods, gamma irradiation can penetrate deep into materials, allowing for the sterilization of pre-packaged and heat-sensitive items. This method is widely used in hospitals and medical manufacturing to maintain sterile environments and prevent infections.

Industrial and Scientific Applications of Gamma Rays

Beyond the realm of medicine, gamma rays have numerous industrial and scientific applications. Their unique properties make them indispensable for non-destructive testing, material analysis, and space exploration.

1. Non-Destructive Testing (NDT)

In industry, gamma rays are used for non-destructive testing to inspect the integrity of materials and structures. By passing gamma rays through objects and capturing the radiation that emerges on the other side, technicians can identify internal defects, such as cracks, voids, or weld flaws. This technique is crucial for ensuring the safety and reliability of critical infrastructure, including pipelines, bridges, and aircraft components. The ability to detect defects without damaging the material under inspection makes gamma-ray NDT a valuable tool in quality control and maintenance.

2. Material Analysis: Gamma-Ray Spectroscopy

Gamma-ray spectroscopy is a powerful analytical technique used to study the composition and properties of materials. By measuring the energy and intensity of gamma rays emitted by a sample, scientists can identify the elements present and determine their concentrations. This method is widely used in fields such as geology, environmental science, and nuclear physics. For example, gamma-ray spectroscopy helps geologists analyze rock samples for trace elements, while nuclear scientists use it to study radioactive decay processes.

3. Space Exploration and Astrophysics

Gamma rays provide a window into some of the most energetic and mysterious phenomena in the universe. Gamma-ray telescopes, such as the Fermi Gamma-ray Space Telescope, have revolutionized our understanding of the cosmos by detecting gamma rays emitted by black holes, neutron stars, and gamma-ray bursts (GRBs). GRBs are the most powerful explosions in the universe, releasing more energy in a few seconds than the

Sun will emit in its entire lifetime. Studying these events has revealed insights into the formation of black holes and the behavior of matter under extreme conditions.

Gamma-ray astronomy has also contributed to our understanding of dark matter and the origins of cosmic rays. By observing gamma rays from distant galaxies and cosmic phenomena, astrophysicists continue to unravel the mysteries of the universe, shedding light on the fundamental forces that shape our existence.

The Dangers of Gamma Rays

While gamma rays have numerous beneficial applications, their high energy and penetrating power make them extremely hazardous if not properly managed. Exposure to gamma radiation can have severe consequences for human health, including acute radiation sickness, an increased risk of cancer, and genetic damage. The dangers of gamma rays necessitate strict safety protocols in both medical and industrial settings.

Health Risks: Gamma rays can penetrate deep into the body, ionizing atoms and molecules in tissues and organs. This ionization can damage DNA and other critical cellular components, potentially leading to cancer or other long-term health effects. Acute exposure to high doses of gamma radiation can cause immediate and life-threatening damage, including burns, organ failure, and radiation poisoning.

Radiation Shielding: Protecting against gamma rays requires the use of dense materials, such as lead or concrete, which can absorb or scatter the radiation. In medical facilities, radiation therapy rooms are equipped with thick walls and lead-lined doors to contain gamma rays and prevent exposure to staff and patients. Industrial workers who handle gamma-ray sources must wear protective clothing and use shielding equipment to minimize their risk of exposure.

Regulation and Safety Standards: Governments and regulatory bodies have established strict guidelines for the use of gamma rays to ensure public safety. These regulations cover everything from the transportation and storage of radioactive materials to the operation of medical and industrial devices that emit gamma radiation. Compliance with these standards is essential for preventing accidents and protecting both workers and the general public.

Gamma rays represent one of the most potent forces in nature, with the power to heal, explore, and destroy. Their applications in medicine, industry, and science have transformed our understanding of the universe and improved the quality of human life. Yet, the very properties that make gamma rays so valuable also make them dangerous, requiring us to approach their use with caution and respect.

As we continue to explore the potential of gamma rays, new technologies and research will undoubtedly expand their applications and improve our ability to harness their energy safely. From life-saving cancer treatments to the discovery of cosmic phenomena, gamma rays remind us of the incredible possibilities that lie within the invisible forces of the universe.

Our journey through the electromagnetic spectrum is a testament to human ingenuity and the relentless pursuit of knowledge. As we advance, gamma rays will continue to be a source of inspiration and discovery, pushing the boundaries of what is possible and revealing the hidden wonders of the world beyond sight.

6. X-RAYS

Revolutionizing Medical Imaging and Beyond

X-rays, a form of electromagnetic radiation with wavelengths ranging from 0.01 to 10 nanometers, have become a cornerstone of modern science and medicine. Their discovery marked the beginning of a new era, fundamentally transforming our understanding of the human body and the universe. From revolutionizing medical diagnostics to uncovering hidden secrets in art and archaeology, X-rays have a diverse array of applications that continue to shape multiple fields. Despite their incredible benefits, the use of X-rays also comes with risks, necessitating a deep understanding of their properties and the precautions needed to use them safely.

This chapter explores the history of X-ray discovery, the underlying science that makes X-rays so effective, their transformative impact on medicine and industry, and the precautions required to mitigate their potential dangers. Let's delve into the world of X-rays and discover how these invisible rays have revolutionized our world.

The Discovery of X-Rays

The story of X-rays begins with a pivotal moment in scientific history: the accidental discovery by Wilhelm Conrad Roentgen in 1895. Roentgen, a German physicist, was experimenting with cathode rays in a vacuum tube when he noticed a mysterious glow on a nearby fluorescent screen. Intrigued by this phenomenon, he realized that an unknown type of ray was penetrating through solid objects and casting shadows of their interiors onto the screen. Roentgen named these mysterious rays "X-rays," with "X" signifying their unknown nature.

Roentgen's discovery was so groundbreaking that he immediately published his findings, and within a few months, X-rays were being used in hospitals to examine broken bones. In 1901, Roentgen received the first-ever Nobel Prize in Physics for his discovery, a testament to the profound impact of X-rays on both science and society. The potential of X-rays was quickly recognized, and research into their properties and applications expanded rapidly, leading to the development of X-ray imaging technology that we use today.

Understanding the Science of X-Rays

X-rays are a type of electromagnetic radiation, just like visible light, but with much shorter wavelengths and higher energy. They fall between ultraviolet light and gamma rays on the electromagnetic spectrum. The energy carried by X-ray photons enables them to pass through soft tissues but be absorbed by denser materials like bones and metal. This unique ability to penetrate matter makes X-rays ideal for medical imaging and various industrial applications.

X-rays are categorized into two main types based on their energy and penetrative abilities:

Soft X-Rays: These X-rays have relatively lower energy and longer wavelengths, making them suitable for imaging soft tissues. They are used in medical and dental X-rays to capture detailed images of internal structures.

Hard X-Rays: With higher energy and shorter wavelengths, hard X-rays can penetrate even dense materials like metals. They are commonly used in industrial radiography, security scanners, and certain medical applications, such as imaging tumors or foreign objects in the body.

The interaction of X-rays with matter is governed by several key physical principles:

Absorption: When X-rays pass through matter, they can be absorbed by atoms, resulting in the attenuation of the X-ray beam. The extent of absorption depends on the density and atomic number of the material. Bones, for instance, absorb more X-rays than soft tissues, creating a contrast that allows for clear imaging.

Scattering: X-rays can also be scattered when they collide with atoms, changing direction and energy. This phenomenon is a key consideration in medical imaging and radiation safety.

Ionization: X-rays are a form of ionizing radiation, meaning they have enough energy to remove electrons from atoms and create ions. While this property is beneficial for certain applications, such as cancer treatment, it also poses health risks, as ionization can damage living tissues and DNA.

Medical Imaging: The Transformative Impact of X-Rays

The most well-known and impactful use of X-rays is in medical imaging, a field that has revolutionized the way doctors diagnose and treat diseases. X-ray imaging has become a fundamental diagnostic tool, enabling healthcare professionals to visualize the internal structures of the body without invasive surgery. Here's how X-rays have transformed medical practice:

1. Diagnostic Radiography

Diagnostic radiography is the most common use of X-rays in medicine. When an X-ray beam is directed through the body, different tissues absorb the rays to varying degrees. Bones, which are dense and rich in calcium, absorb more X-rays and appear white on an X-ray film. Soft tissues, such as muscles and organs, absorb fewer X-rays and appear in shades of gray, while air-filled spaces, like the lungs, appear black.

Diagnostic radiography is used to detect a wide range of conditions, including:

Bone Fractures: X-rays are invaluable for identifying and assessing bone fractures. Orthopedic surgeons rely on X-ray images to guide treatment plans and monitor the healing process.

Chest X-Rays: These are used to diagnose conditions such as pneumonia, lung cancer, and heart enlargement. Chest X-rays are also essential for assessing the placement of medical devices, such as pacemakers and endotracheal tubes.

Dental X-Rays: Dentists use X-rays to examine the teeth, gums, and jawbones. They help in identifying cavities, impacted teeth, and periodontal disease.

2. Fluoroscopy

Fluoroscopy is a special type of X-ray imaging that provides real-time moving images of the internal structures of the body. It is commonly used in procedures such as:

Angiography: This technique visualizes blood vessels and assesses blood flow, often used to diagnose and treat cardiovascular conditions. A contrast dye is injected into the bloodstream, and fluoroscopy captures images of the dye as it travels through the arteries and veins.

Gastrointestinal Studies: Fluoroscopy is used to examine the digestive tract, such as during a barium swallow study, where the patient ingests a contrast agent to highlight the esophagus, stomach, and intestines.

Orthopedic Surgery: Surgeons use fluoroscopy to guide the placement of implants, screws, and other devices during procedures.

Fluoroscopy has the advantage of allowing physicians to observe the movement of internal structures in real time, making it a critical tool for both diagnosis and intervention.

3. Computed Tomography (CT) Scans

Computed Tomography (CT), also known as CAT scans, is a sophisticated imaging technique that uses X-rays to create cross-sectional images of the body. A CT scanner rotates around the patient, capturing multiple X-ray images from different angles. These images are then processed by a computer to generate detailed 3D images of organs, bones, and tissues.

CT scans provide unparalleled clarity and detail, making them indispensable for diagnosing conditions such as:

Cancer: CT scans are used to detect tumors, monitor their growth, and guide biopsies and radiation therapy.

Trauma: In emergency medicine, CT scans are crucial for assessing internal injuries, such as bleeding, organ damage, or fractures.

Neurological Disorders: CT imaging of the brain helps diagnose strokes, brain tumors, and traumatic brain injuries.

The ability of CT scans to provide detailed images of complex anatomical structures has made them a gold standard in medical diagnostics.

Beyond Medicine: The Diverse Applications of X-Rays

While medical imaging remains the most prominent use of X-rays, their applications extend far beyond healthcare. X-rays are employed in various industries, scientific research, and even art restoration.

1. Industrial Radiography

In industry, X-rays are used for non-destructive testing (NDT) to inspect the integrity of materials and structures. This technique is essential for ensuring the safety and reliability of critical components, such as:

Pipelines: X-ray imaging is used to detect cracks, corrosion, or weld defects in pipelines that transport oil and gas.

Aerospace: Aircraft manufacturers use X-rays to examine the structural integrity of airplane parts, ensuring they meet strict safety standards.

Construction: X-rays help identify defects in concrete structures, such as bridges and dams, without causing damage.

Industrial radiography plays a crucial role in maintaining the safety and functionality of infrastructure, making it a vital tool for engineers and quality control specialists.

2. Security Screening

X-rays are widely used in security screening to inspect luggage and cargo at airports, seaports, and border crossings. X-ray scanners can detect weapons, explosives, and contraband by revealing the contents of packages and bags. Advanced security systems use dual-energy X-ray technology to differentiate between organic and inorganic materials, enhancing the accuracy of threat detection.

Security screening is a critical component of modern safety protocols, ensuring the protection of travelers and the prevention of smuggling and terrorism.

3. Scientific Research and Crystallography

X-rays have made significant contributions to scientific research, particularly in the field of X-ray crystallography. This technique is used to determine the atomic and molecular structure of crystals. When X-rays are directed at a crystalline material, they are diffracted in specific patterns that can be analyzed to reveal the arrangement of atoms.

X-ray crystallography has led to groundbreaking discoveries, including:

The Structure of DNA: The double-helix structure of DNA was revealed through the work of Rosalind Franklin and her use of X-ray diffraction.

Drug Development: Pharmaceutical researchers use X-ray crystallography to study the structure of proteins and design drugs that target specific molecular pathways.

Material Science: Scientists use X-rays to analyze the properties of new materials, such as superconductors and nanomaterials.

The precision and power of X-ray crystallography have made it an essential tool for advancing our understanding of the natural world.

4. Art and Archaeology

X-rays have also found applications in the world of art and archaeology, where they are used to study and preserve cultural heritage. X-ray imaging can reveal hidden layers of paint, underlying sketches, and structural damage in artwork without harming the original piece. Archaeologists use X-rays to examine the contents of ancient artifacts, such as mummies or sealed containers, providing valuable insights into history and culture.

The Risks and Safety Precautions of X-Rays

Despite their immense benefits, X-rays are a form of ionizing radiation and pose health risks if not used carefully. Prolonged or excessive exposure to X-rays can damage tissues and increase the risk of cancer. Therefore, strict safety measures are in place to protect both patients and healthcare workers.

Protective Equipment: Lead aprons, thyroid shields, and lead glasses are used to shield patients and medical personnel from unnecessary radiation exposure.

Radiation Dose Management: Medical professionals use the lowest possible dose of X-rays to achieve the necessary diagnostic results. Advances in imaging technology have made it possible to reduce radiation exposure significantly.

Regulation and Monitoring: Facilities that use X-ray equipment are subject to stringent regulations and regular inspections. Radiographers and technicians are trained to operate X-ray machines safely and minimize exposure.

X-rays have truly revolutionized the world, from medicine and industry to science and art. Their ability to penetrate matter and reveal hidden structures has made them indispensable tools across multiple fields. As technology continues to evolve, new applications and innovations in X-ray technology will undoubtedly emerge, further expanding our understanding of the invisible forces that shape our reality.

The journey of X-rays, from their accidental discovery to their widespread use today, is a testament to human curiosity and ingenuity. While we must continue to respect the potential dangers of X-rays and use them responsibly, their contributions to society are immeasurable. X-rays have opened a window into the unseen, allowing us to diagnose diseases, ensure the safety of infrastructure, and unlock the secrets of the universe.

7. ULTRAVIOLET RAYS

The Light We Cannot See and Its Many Faces

Ultraviolet (UV) rays, often shrouded in a veil of mystery, are part of the electromagnetic spectrum, occupying a range just beyond the violet end of visible light. While we cannot see UV rays with the naked eye, their presence and effects are all around us, influencing everything from human health to environmental systems and modern technology. UV radiation plays a dual role in our world: it is both a force for good and a potential hazard, impacting the fields of medicine, astronomy, and even daily human activities.

This chapter delves deep into the science behind UV rays, their applications across various fields, the benefits and risks they pose, and the measures we must take to harness their potential safely. By exploring UV rays in all their complexity, we can better understand and appreciate this powerful, invisible component of sunlight.

The Discovery and Nature of Ultraviolet Rays

The journey of understanding ultraviolet rays began in the early 19th century with the work of Johann Wilhelm Ritter, a German physicist. In 1801, Ritter discovered the existence of invisible rays beyond the violet end of the visible spectrum while conducting experiments with silver chloride, a compound that darkened upon exposure to sunlight. He found that this darkening effect was more pronounced beyond the violet end of the spectrum, indicating the presence of a new form of light, which he called "chemical rays" or "oxidizing rays." These rays were later named "ultraviolet" due to their position beyond violet light.

UV rays, like all forms of electromagnetic radiation, travel in waves and are characterized by their wavelength and energy. The UV spectrum is typically divided into three main categories:

UVA (320-400 nm): Known as "long-wave" UV, UVA rays have the longest wavelengths and the least energy among UV rays. They account for approximately 95% of the UV radiation that reaches the Earth's surface. UVA rays penetrate deep into the skin and are responsible for long-term skin damage, premature aging, and some forms of skin cancer.

UVB (280-320 nm): These "medium-wave" rays have more energy than UVA rays and are partially absorbed by the Earth's ozone layer. UVB rays are responsible for causing sunburn and play a key role in the development of skin cancer. They also contribute to the production of vitamin D in the skin, which is essential for bone health.

UVC (100-280 nm): The most energetic and potentially harmful form of UV radiation, UVC rays are almost entirely absorbed by the Earth's atmosphere, specifically the ozone layer. However, UVC rays can be generated artificially and are used in germicidal lamps and sterilization processes.

Understanding the different types of UV rays and their characteristics has paved the way for harnessing their potential in various fields while also recognizing the need to protect ourselves from their harmful effects.

The Science Behind Ultraviolet Radiation

The interaction of UV radiation with matter is governed by the principles of energy absorption and molecular excitation. When UV rays strike a surface, their energy can be absorbed by the atoms and molecules of that material, leading to various chemical and biological effects. The extent of these effects depends on the energy of the UV rays and the properties of the material they interact with.

Ionization and Excitation: UV radiation has enough energy to excite electrons in atoms and molecules, sometimes causing them to break free from their atomic bonds. This process, known as ionization, can lead to the formation of free radicals—highly reactive molecules that can damage cells and DNA. The ability of UV rays to ionize matter is a key factor in their use in sterilization and their role in causing skin damage and cancer.

Photochemical Reactions: UV rays can initiate photochemical reactions, which are chemical changes that occur when molecules absorb light energy. These reactions are essential in processes like photosynthesis in plants and the synthesis of vitamin D in human skin. However, they can also lead to the degradation of materials, such as the fading of colors in fabrics and the breakdown of plastics.

Fluorescence and Phosphorescence: Some materials can absorb UV light and re-emit it as visible light, a phenomenon known as fluorescence. This property is widely used in various applications, from forensic science to entertainment.

The Beneficial Applications of Ultraviolet Rays

Despite their potential dangers, UV rays have been harnessed for numerous beneficial applications across various fields, from healthcare to environmental science and technology.

1. Medical Applications

UV radiation has long been used in the medical field for both diagnostic and therapeutic purposes. Some of the key medical applications of UV rays include:

Phototherapy: UVB rays are used in phototherapy to treat skin conditions such as psoriasis, eczema, and vitiligo. Exposure to controlled doses of UVB light can help reduce inflammation, slow down the overproduction of skin cells, and repigment the skin.

Sterilization and Disinfection: UVC rays are highly effective at killing bacteria, viruses, and other pathogens. UVC germicidal lamps are widely used in hospitals, laboratories, and water treatment facilities to sterilize equipment and purify water. The COVID-19 pandemic highlighted the importance of UVC technology in disinfecting public spaces and personal protective equipment.

Vitamin D Production: Exposure to UVB rays triggers the synthesis of vitamin D in the skin, which is crucial for bone health, immune function, and overall well-being. In areas with limited sunlight, UV lamps are sometimes used to prevent vitamin D deficiency.

2. Agriculture and Horticulture

UV rays play a significant role in agriculture and horticulture by influencing plant growth and development. Here are some ways UV radiation is used in these fields:

Pest Control: UV light traps are used to attract and capture insects, reducing the need for chemical pesticides. UV light can also disrupt the life cycle of certain pests, helping to control their populations.

Plant Growth and Health: Controlled exposure to UVB radiation can stimulate the production of protective compounds in plants, making them more resistant to pests and diseases. However, excessive UV exposure can damage plant tissues and reduce crop yields, highlighting the need for careful management.

3. Environmental Science and Monitoring

UV radiation is used in environmental science to monitor and study various natural processes. For example:

Water Purification: UVC lamps are used in water treatment systems to kill harmful microorganisms and ensure safe drinking water. This method is effective and environmentally friendly, as it does not require the use of chemicals.

Atmospheric Studies: UV sensors are used to measure the concentration of ozone in the atmosphere and monitor the effects of climate change. Understanding the dynamics of UV radiation and ozone depletion is critical for protecting ecosystems and human health.

4. Industrial and Technological Applications

UV rays have numerous applications in industry and technology, ranging from manufacturing to consumer products. Some notable examples include:

UV Curing: In manufacturing, UV light is used to cure or harden coatings, adhesives, and inks. This process is faster and more energy-efficient than traditional heat-based methods and is used in industries such as printing, automotive, and electronics.

Forensic Science: UV light is a valuable tool in forensic investigations. It can reveal bodily fluids, fingerprints, and other evidence that are invisible under normal lighting conditions. UV lamps are also used to detect counterfeit currency and documents.

Consumer Products: UV filters are commonly used in sunglasses and camera lenses to protect against harmful UV rays. UV nail lamps are used in the beauty industry to cure gel nail polish, providing a long-lasting finish.

The Perils of Ultraviolet Radiation

While UV rays have numerous beneficial applications, they also pose significant risks to human health and the environment. Understanding these risks is crucial for developing effective protective measures.

1. Health Risks

Skin Damage: Prolonged exposure to UV radiation can cause acute and chronic damage to the skin. Sunburn, characterized by redness and inflammation, is a common short-term effect of excessive UV exposure. Over time, UV rays can accelerate skin aging, leading to wrinkles, loss of elasticity, and age spots.

Skin Cancer: UV radiation is a major risk factor for skin cancer, including basal cell carcinoma, squamous cell carcinoma, and melanoma. Melanoma, the most aggressive form of skin cancer, can be life-threatening if not detected and treated early. The risk of skin cancer is higher for individuals with fair skin, a history of sunburns, or a family history of the disease.

Eye Damage: UV rays can also damage the eyes, leading to conditions such as photokeratitis (a painful, temporary burn to the cornea), cataracts, and macular degeneration. Wearing sunglasses with UV protection is essential to prevent these harmful effects.

Immune System Suppression: Excessive UV exposure can weaken the immune system, making the body more susceptible to infections and reducing its ability to fight off diseases.

2. Environmental Impact

Ozone Layer Depletion: The ozone layer in the Earth's stratosphere acts as a protective shield, absorbing most of the sun's harmful UVC and UVB radiation. Human activities, such as the release of chlorofluorocarbons (CFCs), have led to the depletion of the ozone layer, increasing the amount of UV radiation that reaches the Earth's surface. This has serious consequences for ecosystems, including increased skin cancer rates in humans and damage to marine life and crops.

Effects on Wildlife: UV radiation can affect the health and behavior of animals, particularly those that live in or near water. Amphibians, for example, are highly sensitive to UV rays, and increased exposure can lead to deformities, reduced reproductive success, and population declines.

Protecting Ourselves from Ultraviolet Radiation

Given the potential dangers of UV radiation, it is essential to take precautions to minimize exposure and protect ourselves from its harmful effects. Here are some practical measures:

Use Sunscreen: Applying broad-spectrum sunscreen with a high SPF (sun protection factor) can help protect the skin from both UVA and UVB rays. Sunscreen should be applied generously and reapplied every two hours, especially after swimming or sweating.

Wear Protective Clothing: Clothing made from tightly woven fabrics can provide a physical barrier against UV rays. Wide-brimmed hats and UV-protective sunglasses are also important for shielding the face and eyes from the sun.

Seek Shade: Avoiding direct sunlight during peak hours (10 a.m. to 4 p.m.) can significantly reduce the risk of UV exposure. Seeking shade under trees, umbrellas, or canopies can provide relief from the sun's rays.

Be Cautious with Artificial UV Sources: Tanning beds and sunlamps emit concentrated UV radiation and should be used with caution. The World Health Organization classifies tanning devices as a known carcinogen, and their use is strongly discouraged, especially for young people.

Ultraviolet rays, though invisible to the human eye, have a profound impact on our world. They are both a source of life and a potential threat, playing a crucial role in processes such as vitamin D synthesis and the disinfection of water, while also posing risks to skin health and contributing to environmental challenges.

The dual nature of UV radiation underscores the importance of continued research and innovation. By harnessing the benefits of UV rays in fields such as medicine, agriculture, and environmental science, while also implementing protective measures to minimize their harmful effects, we can create a safer and healthier world. As we deepen our understanding of ultraviolet radiation, we open new possibilities for using this invisible force to improve our lives and safeguard our planet.

In the end, the story of UV rays is a reminder of the delicate balance between harnessing nature's power and respecting its potential dangers. The light we cannot see continues to shape our world in ways that are both fascinating and vital, illuminating the invisible connections between science, health, and the environment.

8. INFRARED RAYS

From Heat Vision to Cutting-Edge Technology

Infrared (IR) rays, part of the electromagnetic spectrum just beyond the visible light range, have transformed the way we perceive the world and harness energy. Invisible to the naked eye, infrared radiation is often experienced as heat, making it one of nature's most essential and ubiquitous forces. From the warmth of the sun to the technological marvels that define modern science, infrared rays have found applications across numerous fields, from medical diagnostics to military operations.

This chapter dives into the nature of infrared rays, exploring their discovery, characteristics, and the extraordinary innovations they have sparked in the fields of science, medicine, industry, and even everyday life.

The Discovery and Nature of Infrared Rays

The journey of infrared rays began in the early 19th century, thanks to the pioneering work of Sir William Herschel, an astronomer best known for discovering the planet Uranus. In 1800, while experimenting with a prism to study sunlight, Herschel aimed to measure the temperature of different colors of light. To his surprise, he found that the region beyond the red end of the visible spectrum, where no light was visible, was warmer than any of the colors he had measured. This discovery marked the beginning of our understanding of infrared radiation.

Infrared rays are characterized by their wavelengths, which are longer than those of visible light but shorter than those of microwaves. The infrared spectrum is typically divided into three categories:

Near-Infrared (NIR) (0.7-1.5 micrometers): The closest to visible light, near-infrared rays are used in applications like remote controls, fiber optic communications, and infrared spectroscopy.

Mid-Infrared (MIR) (1.5-5 micrometers): These rays are often associated with thermal imaging and can detect heat emitted by objects. They are widely used in medical diagnostics, industrial monitoring, and environmental sensing.

Far-Infrared (FIR) (5-1,000 micrometers): This part of the spectrum is primarily associated with heat and is used in applications like climate studies, heating systems, and astrophysical observations.

The diversity of infrared wavelengths gives rise to an extensive range of applications, each capitalizing on the unique properties of infrared rays.

The Science Behind Infrared Radiation

Infrared radiation is fundamentally about heat and energy transfer. When an object absorbs energy, whether from sunlight or another source, it emits infrared radiation in the form of heat. This principle is at the heart of countless natural and technological processes, making infrared rays an essential part of our world.

Heat Transfer and Blackbody Radiation: Every object with a temperature above absolute zero (-273.15°C or -459.67°F) emits infrared radiation. The amount and wavelength of the radiation depend on the object's temperature. This phenomenon, known as blackbody radiation, explains why warmer objects emit more infrared energy and why infrared cameras can detect heat differences between objects.

Infrared Absorption and Emission: Different materials absorb and emit infrared radiation at different rates. For example, metals reflect most infrared rays, while water and organic tissues absorb them readily. This property is crucial for applications like thermal imaging, where the goal is to detect variations in heat.

Interaction with Matter: Infrared radiation can penetrate various materials, such as fog, smoke, and even human tissue to some extent. This penetrating ability is utilized in medical and industrial imaging to see through obstacles that would be opaque to visible light.

Everyday Encounters with Infrared Rays

Although we cannot see infrared rays, we experience their effects daily. The warmth of the sun, the heat from a fireplace, and even the temperature of our own bodies are all forms of infrared radiation. This omnipresent energy has paved the way for many technologies we now take for granted.

1. Remote Controls and Communication

One of the most familiar uses of infrared technology is in remote controls for televisions, air conditioners, and other electronic devices. These remotes emit pulses of near-infrared light, which are detected by sensors in the device they control. The pulses encode binary data, allowing the device to interpret commands like changing the channel or adjusting the volume.

Infrared communication is also widely used in short-range data transfer. Devices like smartphones and laptops often use infrared beams for file sharing, though this technology has largely been supplanted by Bluetooth and Wi-Fi.

2. Infrared Thermometers

The rise of infrared thermometers, especially during the COVID-19 pandemic, highlighted the importance of infrared technology in health and safety. These devices measure the infrared radiation emitted by a person's skin, providing a quick and non-invasive way to assess body temperature. Infrared thermometers are now commonly used in hospitals, airports, and public places to screen for fever.

Infrared Rays in Medicine

Infrared technology has revolutionized the medical field, offering new ways to diagnose and treat various conditions. From imaging techniques to therapeutic applications, infrared rays have become a vital tool for healthcare professionals.

1. Thermal Imaging and Diagnosis

Thermal imaging, or thermography, uses infrared cameras to detect heat patterns and blood flow in body tissues. This technology is particularly useful for identifying inflammation, infections, and circulatory issues. For example, breast thermography is sometimes used as a complementary tool to detect abnormalities that may indicate breast cancer.

Thermal imaging is also used in sports medicine to monitor muscle and joint conditions. By analyzing heat distribution, doctors can identify injuries and plan appropriate treatments. Additionally, thermography has applications in detecting fevers, vascular disorders, and even monitoring the progression of chronic conditions like diabetes.

2. Infrared Therapy

Infrared therapy, which uses infrared lamps or lasers to generate heat, has gained popularity for pain relief and tissue healing. The heat penetrates deep into muscles and tissues, increasing blood flow and reducing inflammation. This therapy is used to treat conditions like arthritis, muscle pain, and sports injuries.

Far-infrared saunas are another example of infrared therapy. These saunas use FIR heaters to raise the body's core temperature, promoting relaxation, detoxification, and improved circulation. Studies have shown that regular use of infrared saunas can provide health benefits, including reduced blood pressure and improved cardiovascular health.

3. Surgical and Dental Applications

Infrared lasers are used in various surgical procedures, including eye surgeries like LASIK and dental treatments. These lasers offer precision and minimal tissue damage compared to traditional surgical tools, reducing recovery times for patients. In dentistry, infrared lasers are used for cavity detection, gum reshaping, and root canal treatments.

Industrial and Scientific Applications

Infrared rays have found numerous applications in industry and science, driving innovation and efficiency across multiple sectors.

1. Thermal Imaging for Inspection and Maintenance

Thermal imaging cameras are widely used in industrial settings to monitor equipment and detect potential issues. These cameras can identify overheating machinery, electrical faults, and leaks in pipelines by detecting temperature anomalies. This non-contact method of inspection improves safety and reduces the risk of equipment failure.

In the construction industry, infrared imaging is used to inspect buildings for energy efficiency. By detecting areas where heat is lost or gained, builders can improve insulation and reduce energy consumption. This technology is also employed in firefighting to locate hot spots and trapped individuals in burning buildings.

2. Environmental Monitoring

Infrared technology plays a crucial role in monitoring the environment and studying climate change. Satellites equipped with infrared sensors can measure sea surface temperatures, monitor deforestation, and track the movement of pollutants. These observations help scientists understand the impact of climate change on ecosystems and develop strategies to mitigate its effects.

Infrared spectroscopy is another valuable tool for environmental analysis. It is used to identify chemical compounds in air, water, and soil samples, providing insights into pollution levels and the presence of hazardous substances.

3. Astronomy and Space Exploration

Infrared astronomy has opened new windows into the universe, allowing scientists to observe celestial objects that are invisible in visible light. Stars, planets, and galaxies often emit infrared radiation, which can penetrate dust clouds that obscure them in the visible spectrum. Infrared telescopes, such as the James Webb Space Telescope (JWST), have revolutionized our understanding of the cosmos, revealing the formation of stars and the atmospheres of distant exoplanets.

Infrared technology is also used in space exploration to study the surface of planets and moons. For instance, the Mars Curiosity Rover is equipped with an infrared spectrometer to analyze the composition of Martian rocks and soil, helping scientists search for signs of past or present life.

Military and Security Applications

Infrared technology has become an integral part of modern military and security operations. Its ability to detect heat signatures makes it invaluable for surveillance, navigation, and target acquisition.

1. Night Vision and Thermal Imaging

One of the most well-known applications of infrared technology is in night vision devices. These devices amplify available light or detect heat emitted by objects, allowing soldiers and security personnel to see in complete darkness. Thermal imaging cameras are used to detect enemy movements, even through fog, smoke, or camouflage.

Night vision goggles, often used by the military, have also found applications in civilian law enforcement and search-and-rescue missions. Thermal imaging is used to locate missing persons, track suspects, and monitor wildlife.

2. Guided Missiles and Targeting Systems

Infrared technology is used in guided missiles and targeting systems to track the heat signatures of enemy vehicles, aircraft, or infrastructure. Infrared homing missiles, for example, can lock onto the heat emitted by an aircraft's engines, making them highly effective in combat. Infrared cameras are also used in drones for reconnaissance and surveillance missions.

3. Border Security and Surveillance

Infrared cameras are deployed along national borders to monitor activity and prevent illegal crossings. These cameras can detect movement in low-light conditions, providing security personnel with real-time information about potential threats. Infrared technology is also used in perimeter security systems for critical infrastructure, such as power plants and airports.

Consumer and Everyday Applications

Infrared technology has made its way into our daily lives, making many tasks more convenient and efficient.

1. Smartphones and Wearable Devices

Many modern smartphones are equipped with infrared sensors for features like facial recognition and remote control functionality. These sensors can detect heat patterns, enabling secure and contactless authentication. Infrared sensors are also used in wearable fitness trackers to monitor heart rate and body temperature.

2. Home Automation

Infrared technology is a cornerstone of smart home automation. Devices like smart thermostats use infrared sensors to detect when people are present in a room, adjusting the temperature accordingly. Infrared motion sensors are also used in security systems and automated lighting, making homes more energy-efficient and secure.

3. Automotive Safety

Infrared technology has enhanced vehicle safety with features like adaptive cruise control and night vision systems. These systems use infrared sensors to detect obstacles, pedestrians, and other vehicles, even in low visibility conditions. Infrared sensors are also used in self-driving cars to navigate and avoid collisions.

The Future of Infrared Technology

As technology continues to advance, the potential applications of infrared rays are expanding. Researchers are exploring new ways to harness infrared radiation for renewable energy, such as developing materials that can convert heat into electricity. Infrared spectroscopy is being refined for more precise medical diagnostics and environmental monitoring.

In the field of communication, scientists are working on developing faster and more secure data transmission methods using infrared light. Infrared quantum communication, for example, holds promise for creating unbreakable encryption, ensuring data security in an increasingly connected world.

ConclusionInfrared rays, though invisible to our eyes, are a powerful and versatile force that has shaped modern science and technology. From the warmth of a sunny day to the precision of medical imaging, infrared radiation permeates every aspect of our lives, often without us even realizing it. Its applications in medicine, industry, security, and space exploration continue to evolve, pushing the boundaries of what is possible.

As we continue to harness the power of infrared rays, the challenge lies in using this invisible energy responsibly and innovatively. The story of infrared technology is far from over, and the future promises even more groundbreaking discoveries that will further illuminate our understanding of the world—and the universe—around us.

9. MICROWAVES

Connecting the World and Exploring the Universe

Microwaves, a form of electromagnetic radiation with wavelengths ranging from one millimeter to one meter, are an integral part of our modern existence. They are used to cook our food, power our wireless communications, and even help astronomers peer into the distant reaches of the universe. As a bridge between radio waves and infrared rays on the electromagnetic spectrum, microwaves occupy a unique space, offering applications that span across everyday conveniences and cutting-edge scientific research.

In this chapter, we'll explore the fundamental properties of microwaves, their technological uses, their role in understanding the cosmos, and how they continue to influence the fabric of human society and scientific advancement.

The Science of Microwaves

Microwaves are characterized by their frequencies, which range from 300 MHz (0.3 GHz) to 300 GHz. This range includes a broad spectrum of applications, from radar and satellite communications to microwave ovens and scientific measurements.

1. Properties of Microwaves

Microwaves exhibit several unique properties that make them exceptionally useful across different fields:

Penetration: Microwaves can penetrate clouds, smoke, and certain building materials, allowing them to be used for communication and remote sensing, even under challenging conditions.

Reflection: They are easily reflected by metals, making them ideal for radar systems, which use metal reflections to detect and map objects.

Absorption: Microwaves are absorbed by water, fats, and sugars, which is why they are effective for heating food. This property is harnessed in microwave ovens, where the absorbed energy vibrates water molecules, generating heat.

These properties make microwaves versatile for both everyday applications and complex scientific investigations.

Microwaves in Daily Life: From Kitchens to Communications

Microwaves have become an indispensable part of everyday life, from the appliances in our kitchens to the phones in our pockets. Let's explore two major areas where microwaves have had a transformative impact: domestic appliances and wireless communication.

1. Microwave Ovens: The Revolution in Cooking

The invention of the microwave oven was an accidental yet revolutionary breakthrough in the world of domestic cooking. In the 1940s, engineer Percy Spencer discovered that microwaves could cook food when he noticed that a candy bar in his pocket melted while he was working with a magnetron, a device that produces microwaves.

A microwave oven works by emitting microwaves that are absorbed by the water, fat, and sugar molecules in food. As these molecules absorb energy, they vibrate rapidly, generating heat and cooking the food. Unlike traditional ovens, which cook from the outside in, microwaves penetrate food and heat it evenly, significantly reducing cooking time.

While microwave ovens have simplified meal preparation for millions, they have also raised questions about food safety and nutritional impact. However, studies have shown that microwaving is generally safe and often preserves nutrients better than boiling or frying, as it uses less heat and cooking time.

2. Wireless Communication: The Backbone of Connectivity

Perhaps the most significant and far-reaching application of microwaves is in the realm of wireless communication. From cell phones and Wi-Fi to satellite and GPS, microwaves are the invisible force that keeps our world connected.

Cellular Networks: Mobile phones rely on microwave frequencies to transmit and receive signals. These signals are carried between cell towers and mobile devices, enabling instant communication over vast distances.

Wi-Fi: Wireless internet, or Wi-Fi, uses microwaves in the 2.4 GHz and 5 GHz frequency bands to connect devices to the internet. These frequencies are chosen because they can efficiently transmit data without causing significant interference.

Bluetooth: Bluetooth technology, used for short-range communication between devices, also operates in the microwave frequency range. It allows for the seamless exchange of data between gadgets like smartphones, headphones, and smartwatches.

Satellite Communication: Satellites use microwaves to relay television, radio, and internet signals across the globe. These signals can travel vast distances in space, connecting even the most remote areas.

Microwave communication has reshaped our daily lives, enabling everything from video calls to global positioning systems. As the world becomes increasingly interconnected, microwave technology continues to evolve, paving the way for faster and more reliable communication networks.

Microwaves in Industry and Medicine

Beyond cooking and communication, microwaves play a critical role in various industrial and medical applications, demonstrating their wide-reaching influence on modern technology.

1. Industrial Applications

Microwaves are used in a range of industrial processes, including material drying, chemical synthesis, and even quality control.

Microwave Drying: Traditional drying methods, such as air drying or oven drying, can be inefficient and time-consuming. Microwave drying, on the other hand, provides rapid and uniform drying of materials, making it useful in the food, pharmaceutical, and textile industries.

Chemical Synthesis: In the field of chemistry, microwaves are used to accelerate chemical reactions. This technique, known as microwave-assisted synthesis, reduces reaction times and energy consumption, making it a greener alternative to conventional methods.

Nondestructive Testing: Microwaves are used for the nondestructive testing of materials, allowing engineers to inspect the integrity of structures without causing damage. This is particularly valuable in industries like aerospace and construction.

2. Medical Applications

Microwave technology has found its way into healthcare, where it is used for diagnostic and therapeutic purposes.

Microwave Imaging: Medical imaging techniques, such as microwave tomography, use microwaves to create images of the human body. This non-invasive method is being explored for breast cancer detection, as microwaves can differentiate between healthy and cancerous tissue based on their dielectric properties.

Hyperthermia Treatment: Microwaves are used in hyperthermia therapy, a cancer treatment that involves raising the temperature of tumor tissues to damage and kill cancer cells. This method is often used in combination with radiation therapy or chemotherapy to improve treatment outcomes.

Microwave Ablation: Microwave ablation is a minimally invasive procedure used to treat tumors. By using microwaves to generate heat, doctors can destroy cancerous tissue without the need for major surgery. This technique is commonly used for liver, lung, and kidney cancers.

The use of microwaves in medicine continues to expand, offering new possibilities for diagnosing and treating various conditions with greater precision and efficiency.

Radar Technology: Seeing with Microwaves

Radar (Radio Detection and Ranging) is one of the most well-known applications of microwaves. Developed during World War II, radar technology uses microwave signals to detect and locate objects, making it invaluable for aviation, maritime navigation, weather forecasting, and military operations.

1. How Radar Works

Radar systems transmit microwave pulses, which travel through the air and bounce off objects. The radar receiver then detects the reflected signals, measuring the time it takes for the pulses to return. By analyzing this data, radar systems can determine the distance, speed, and direction of an object.

Radar operates in several frequency bands, each suited for different applications:

X-Band Radar: Used in maritime navigation and weather monitoring, X-band radar can detect small objects, such as boats and raindrops.

S-Band Radar: Commonly used in air traffic control and weather surveillance, S-band radar can penetrate precipitation and provide reliable data on storms and aircraft.

L-Band Radar: Used for long-range surveillance and GPS, L-band radar can cover vast areas and is less affected by atmospheric conditions.

Microwaves in Space Exploration

Microwaves are also key to our exploration and understanding of the universe. From cosmic microwave background radiation to space-based communication, these rays are an essential tool for astronomers and space scientists.

1. Cosmic Microwave Background (CMB) Radiation

One of the most significant discoveries in modern cosmology is the cosmic microwave background radiation, a faint glow of microwaves that fills the universe. The CMB is the remnant radiation from the Big Bang, providing a snapshot of the universe when it was just 380,000 years old. It is a crucial piece of evidence supporting the Big Bang theory and has helped scientists understand the early stages of the universe's formation.

In 1965, physicists Arno Penzias and Robert Wilson accidentally discovered the CMB while investigating noise in their microwave antenna. Their discovery earned them the Nobel Prize in Physics and revolutionized our understanding of cosmology.

2. Microwave Astronomy

Microwave astronomy involves studying celestial objects using microwave wavelengths. This field has provided insights into the structure and composition of the universe, including the formation of galaxies and the distribution of dark matter. Telescopes like the Planck Observatory and the Wilkinson Microwave Anisotropy Probe (WMAP) have mapped the CMB with incredible precision, offering clues about the universe's expansion and the distribution of matter.

3. Space-Based Communication

Microwave communication is vital for space exploration, enabling communication between Earth and spacecraft. Deep-space missions, such as those to Mars and beyond, rely on microwave signals to transmit data back to Earth. The Deep Space Network (DSN), managed by NASA, uses large parabolic antennas to receive these signals, ensuring that space probes can send images and scientific data across vast cosmic distances.

Microwave technology also supports the operation of satellites, which are used for weather forecasting, GPS navigation, and global communications. These satellites use microwave frequencies to transmit data, making them crucial for both scientific research and everyday life.

The Role of Microwaves in Weather Forecasting

Accurate weather forecasting is essential for agriculture, aviation, disaster management, and daily life. Microwaves play a vital role in this process, as they are used in weather radar systems and satellite imaging.

1. Weather Radar

Weather radar systems use microwaves to detect precipitation, track storms, and measure rainfall intensity. By analyzing the Doppler shift of reflected microwave signals, meteorologists can determine the speed and direction of moving weather systems. This information is critical for issuing severe weather warnings and tracking hurricanes and tornadoes.

2. Satellite Imaging

Microwave sensors on weather satellites provide data on atmospheric conditions, such as temperature, humidity, and cloud cover. These sensors can penetrate clouds and collect information about weather patterns, even in remote areas where ground-based measurements are not possible. This data is used to create accurate weather models and predict climate changes.

The Future of Microwave Technology

As we look to the future, microwave technology continues to evolve, offering new opportunities and challenges. The development of 5G networks promises faster and more reliable internet connections, using high-frequency microwaves to support the ever-growing demand for data. However, this also raises concerns about radiation exposure and the environmental impact of new infrastructure.

In the field of healthcare, microwave-based treatments are being refined to offer more targeted and effective therapies. Researchers are also exploring the use of microwaves in renewable energy, such as wireless power transfer and solar power harvesting.

Space exploration will continue to benefit from advancements in microwave technology, as scientists develop more sensitive instruments to study the universe. The search for extraterrestrial life, for example, may one day be aided by microwave signals, as we listen for signs of communication from other civilizations.

Microwaves are a testament to the power of invisible forces in shaping our world and expanding our understanding of the universe. From cooking dinner to exploring the cosmos, these rays have revolutionized technology, science, and society. As we continue to harness the potential of microwaves, the possibilities are as vast as the universe itself, promising innovations that will connect, heal, and enlighten future generations.

Microwaves are more than just a convenient way to heat food; they are a bridge between the human experience and the mysteries of the cosmos, an invisible force that binds our world in ways we are only beginning to fully appreciate.

10. RADIO WAVES

The Foundation of Global Communication

Radio waves, the longest wavelength within the electromagnetic spectrum, are essential to modern society's infrastructure. Spanning frequencies from about 3 kHz to 300 GHz, radio waves form the backbone of global communication, transmitting information over vast distances and connecting people in real-time across continents. Whether through AM/FM broadcasting, wireless networks, or satellite communication, radio waves have revolutionized the way we communicate, interact, and understand the world around us.

In this article, we will explore the science behind radio waves, the development of radio technology, and their countless applications. From early radio pioneers to the latest advances in wireless communication, radio waves have paved the way for innovations that have transformed modern civilization.

The Science of Radio Waves

Understanding the properties of radio waves requires a look at their place in the electromagnetic spectrum. Radio waves have the longest wavelengths, typically ranging from a few millimeters to several kilometers, and relatively low frequencies compared to other forms of radiation like infrared or X-rays. These characteristics give radio waves a unique ability to travel over great distances and penetrate various obstacles, making them ideal for communication.

Properties of Radio Waves

Long Wavelengths: The long wavelengths of radio waves mean they can travel great distances without significant loss of energy. This property is especially useful for broadcasting and long-distance communication.

Low Frequencies: The lower frequencies of radio waves, compared to microwaves or infrared rays, allow them to pass through solid objects like buildings and trees, making them suitable for urban communication networks.

Reflection and Refraction: Radio waves can reflect off various surfaces, including the ground and the atmosphere. This property enables them to bounce between the Earth's surface and the ionosphere, allowing them to cover vast distances. Longwave and shortwave radio, for example, use this reflection to reach listeners around the world.

Absorption: Lower-frequency radio waves are less likely to be absorbed by materials compared to higher frequencies, which allows for effective signal transmission over long distances.

Together, these properties have allowed radio waves to be adapted for numerous applications, from the first radio broadcasts to the sophisticated satellite networks that power global positioning systems (GPS) and internet services today.

The History of Radio Communication

The journey of radio technology from theoretical physics to global communication infrastructure is filled with groundbreaking discoveries and developments by pioneering scientists and inventors.

Early Theoretical Foundations

In the 19th century, James Clerk Maxwell formulated the theory of electromagnetism, predicting the existence of electromagnetic waves, including radio waves. Maxwell's equations provided the theoretical basis for understanding how electric and magnetic fields propagate through space.

Heinrich Hertz's Experiments

In 1888, Heinrich Hertz became the first scientist to experimentally confirm the existence of radio waves, proving Maxwell's predictions. Hertz's work involved generating and detecting radio waves using simple circuits, demonstrating their properties such as reflection, refraction, and polarization. Today, his name lives on in the unit of frequency, "hertz."

Guglielmo Marconi and Wireless Telegraphy

Italian inventor Guglielmo Marconi is often credited with developing practical wireless communication systems. In 1895, Marconi sent the first wireless telegraph message across a short distance. By 1901, he had successfully transmitted radio signals across the Atlantic Ocean, paving the way for the first transcontinental communication. Marconi's work established radio waves as a reliable means of transmitting information and set the stage for global communication networks.

The Rise of Broadcast Radio

By the 1920s, radio technology had advanced to the point where regular public broadcasts became feasible. In 1920, the first commercial radio station, KDKA in Pittsburgh, Pennsylvania, began broadcasting news, music, and entertainment to the public. The advent of broadcast radio brought people together, providing a shared experience that transcended geographic barriers and establishing radio as a central part of social life.

World War II and the Expansion of Radio Technology

During World War II, radio technology played a crucial role in military communication, intelligence, and radar systems. Advances in radio engineering led to innovations such as frequency modulation (FM), which improved audio quality and reduced interference compared to amplitude modulation (AM).

Post-War Developments and the Birth of Television

The end of World War II marked a period of rapid technological advancement. The rise of television broadcasting and the development of transistors revolutionized the electronics industry, making radios smaller, cheaper, and more accessible. By the 1950s, radio waves were powering both AM/FM radio and television, transforming how people received news and entertainment.

Types of Radio Waves and Their Uses

The versatility of radio waves lies in the different frequency bands, each of which has unique properties and uses. Some of the most significant bands include:

Very Low Frequency (VLF): VLF waves, from 3 to 30 kHz, are used primarily for military communication and submarine communications, as they can penetrate deep underwater and travel long distances.

Low Frequency (LF): LF waves, from 30 to 300 kHz, are used in maritime communication, navigation, and certain emergency broadcast systems.

Medium Frequency (MF): MF waves, from 300 to 3,000 kHz, are best known for AM radio broadcasts. Due to their range and ability to reflect off the ionosphere, they are used to cover large geographic areas.

High Frequency (HF): HF waves, from 3 to 30 MHz, are used for shortwave radio broadcasts and amateur (ham) radio. They can travel long distances by bouncing off the Earth's surface and the ionosphere, making them useful for international broadcasts.

Very High Frequency (VHF): VHF waves, from 30 to 300 MHz, are used in FM radio, television broadcasting, and two-way radio communication. VHF signals generally travel in a line of sight and are used over short to medium distances.

Ultra High Frequency (UHF): UHF waves, from 300 to 3,000 MHz, are used in television broadcasting, mobile phones, GPS, and Wi-Fi. They provide better penetration through obstacles than VHF and are essential for many modern wireless systems.

Microwave Frequencies: Although not traditionally classified as radio waves, microwaves (300 MHz to 300 GHz) are closely related and share many applications in communication, radar, and satellite systems.

Radio Waves in Modern Communication

Radio waves have become the foundation of modern communication, providing the infrastructure for wireless communication systems that span the globe. Some of the most influential uses include:

Broadcast Radio

Traditional AM and FM radio continue to be significant mediums for news, entertainment, and emergency broadcasts. While digital streaming has grown in popularity, radio remains a reliable and accessible source of information, especially in rural and remote areas.

Television Broadcasting

Radio waves in the VHF and UHF bands have long been used to broadcast television signals. Digital television technology has improved signal quality, allowing for high-definition broadcasting and the delivery of multiple channels within the same frequency.

Mobile Phones

Mobile phones rely on radio waves in the UHF and microwave bands to connect to cell towers, enabling voice, text, and data communication. As networks have evolved from 2G to 5G, the demand for higher frequencies and bandwidth has increased, leading to faster data speeds and enhanced connectivity.

Wi-Fi and Bluetooth

Wi-Fi and Bluetooth technology also operate within the radio wave spectrum. Wi-Fi uses the 2.4 GHz and 5 GHz bands, allowing devices to connect to the internet wirelessly. Bluetooth, on the other hand, operates within the 2.4 GHz band and is commonly used for short-range communication between devices such as headphones, keyboards, and smartphones.

Satellite Communication

Satellites orbiting the Earth use radio waves to communicate with ground stations, enabling satellite TV, GPS, and internet services. Satellite communication plays a vital role in connecting remote areas, providing disaster relief communication, and supporting global navigation.

Emergency and Military Communication

Radio waves are essential for military operations, providing secure, reliable communication in remote areas and challenging conditions. Emergency services, including police, fire, and medical personnel, rely on radio communication for real-time coordination and response.

Radio Astronomy: Unlocking the Secrets of the Universe

Radio waves are not only used for communication but also serve as a powerful tool in astronomy. Radio telescopes detect radio emissions from celestial objects, providing insights into phenomena that are invisible to optical telescopes.

Discovery of Pulsars and Quasars

Radio astronomy led to the discovery of pulsars—rapidly spinning neutron stars that emit beams of radio waves—and quasars, which are extremely luminous objects powered by supermassive black holes at the centers of distant galaxies. These discoveries have expanded our understanding of the universe and its extreme environments.

Cosmic Microwave Background Radiation

One of the most significant discoveries in radio astronomy was the cosmic microwave background radiation (CMB), a faint glow left over from the Big Bang. This discovery provided crucial evidence for the Big Bang theory and offered a snapshot of the early universe.

Mapping the Milky Way

Radio telescopes have allowed astronomers to map the structure of the Milky Way galaxy, studying hydrogen emissions that reveal the distribution of gas and the formation of stars. This information helps scientists understand the processes that shape galaxies.

The Search for Extraterrestrial Intelligence (SETI)

The Search for Extraterrestrial Intelligence, or SETI, uses radio telescopes to listen for signals from potential alien civilizations. By scanning the cosmos for radio waves that could indicate technological activity, scientists hope to answer one of humanity's oldest questions: Are we alone in the universe?

The Future of Radio Wave Technology

The potential of radio waves continues to grow as technology advances. Here are some emerging areas where radio waves are expected to play a vital role:

6G Networks

As the next generation of mobile networks, 6G is anticipated to use higher frequencies within the radio and microwave bands. This advancement promises faster data speeds, enhanced connectivity, and the development of new applications like holographic communication and real-time augmented reality.

Quantum Communication

Researchers are exploring the potential for radio waves in quantum communication, a field that promises ultra-secure transmission of information. Quantum radio systems could enable faster and more secure communication networks.

Space-Based Radio Networks

With the increasing interest in space exploration and communication, radio waves will be essential for connecting space stations, satellites, and rovers on distant planets. Space-based radio networks could support missions to Mars and beyond, providing real-time communication with Earth.

Expanding the Internet of Things (IoT)

As the IoT ecosystem expands, radio waves will enable smart devices to communicate seamlessly. This network of connected devices has applications in various sectors, including healthcare, agriculture, and urban planning, where data-driven insights improve efficiency and quality of life.

Radio waves have transformed human society, creating a world where information flows freely, distances shrink, and new opportunities arise. From the first wireless telegraphs to the modern internet, radio waves have been a driving force behind the communication revolution, connecting individuals, communities, and nations in unprecedented ways.

As technology advances, the role of radio waves will only expand, opening new frontiers in science, technology, and exploration. Whether in a rural village receiving news broadcasts or a satellite orbiting a distant planet, radio waves continue to be the foundation of global communication—an invisible thread that binds our world and pushes the boundaries of what is possible.

11. THE PHYSICS OF INVISIBLE RAYS

Understanding Wavelengths, Frequencies, and Energy

Invisible rays, encompassing everything from radio waves to gamma rays, form a vast part of the electromagnetic spectrum and are vital to our understanding of both physics and the universe. While visible light occupies only a small section, the broader spectrum holds waves and particles that play essential roles in communication, medical imaging, nuclear energy, and astronomical observation. Each type of ray within this spectrum has unique properties of wavelength, frequency, and energy, which dictate their behaviors, interactions, and applications.

This chapter explores the physics behind these invisible rays, delving into the fundamental properties that govern their characteristics and examining how these principles allow us to harness them in various technologies. By understanding wavelengths, frequencies, and energy, we gain insight into the remarkable ways invisible rays influence our world.

1. The Electromagnetic Spectrum: A Brief Overview

The electromagnetic spectrum is a continuum of all electromagnetic waves arranged by frequency and wavelength, from the longest radio waves to the shortest gamma rays. These waves propagate as oscillations of electric and magnetic fields, traveling at the speed of light in a vacuum. The spectrum is divided into categories that range from low-energy, low-frequency waves like radio waves to high-energy, high-frequency gamma rays.

Each category of ray has its own properties and applications, which result from variations in wavelength, frequency, and energy. The main regions of the electromagnetic spectrum, ordered from longest wavelength to shortest, include:

Radio Waves: Low-frequency waves used in communication.

Microwaves: Used in radar, satellite communication, and cooking.

Infrared Rays: Detected as heat and used in night-vision technology.

Visible Light: The small part of the spectrum detectable by human eyes.

Ultraviolet Rays: Higher energy than visible light, involved in sunburns and sterilization.

X-Rays: Used in medical imaging.

Gamma Rays: The highest energy waves, produced in nuclear reactions and certain astronomical phenomena.

Each of these rays differs fundamentally due to its wavelength, frequency, and energy. These differences are critical to their behavior and utility in different applications.

2. Wavelength: The Key to Differentiating Rays

Wavelength refers to the distance between two consecutive peaks or troughs in a wave. It is measured in meters, though for different types of rays, the wavelength can vary significantly—from kilometers in radio waves to picometers (trillionths of a meter) in gamma rays. Wavelength has an inverse relationship with frequency; longer wavelengths correspond to lower frequencies and lower energies.

Wavelength Ranges in the Electromagnetic Spectrum

Radio Waves: Wavelengths range from about one millimeter to hundreds of kilometers.

Microwaves: Wavelengths are between one millimeter and 30 centimeters.

Infrared Rays: These waves have wavelengths from 700 nanometers (nm) to one millimeter.

Visible Light: This narrow band has wavelengths from approximately 400 nm (violet) to 700 nm (red).

Ultraviolet Rays: Wavelengths range from about 10 nm to 400 nm.

X-Rays: Wavelengths are between 0.01 nm and 10 nm.

Gamma Rays: These waves have the shortest wavelengths, typically less than 0.01 nm.

Wavelength is essential in determining a ray's interactions with matter. For example, radio waves with long wavelengths can pass through walls and buildings, making them ideal for communication. On the other hand, gamma rays, with their extremely short wavelengths, can penetrate deep into materials, including human tissue, making them useful in medical treatments and imaging but also requiring caution due to their potential to damage cells.

Wavelength in Practical Applications

Understanding wavelength is crucial for applications that rely on wave behavior, such as:

Telecommunication: Radio waves' long wavelengths make them suitable for transmitting signals over large distances.

Medical Imaging: Short-wavelength X-rays can penetrate soft tissue to reveal internal structures like bones.

Astronomy: Different wavelengths are used to observe various cosmic phenomena, from radio telescopes that capture distant galaxies to X-ray telescopes that reveal black holes.

The range of wavelengths across the spectrum enables us to utilize invisible rays in ways that cater to their unique interactions with the environment.

3. Frequency: The Pace of Oscillation

Frequency measures the number of wave cycles that pass a point in one second and is measured in hertz (Hz). The frequency of an electromagnetic wave determines how much energy it carries and how it interacts with matter. As wavelength decreases, frequency increases, resulting in higher energy levels.

The Relationship between Frequency and Wave Behavior

Frequency impacts how waves interact with materials:

Low-Frequency Waves (Radio): Radio waves have low frequencies, which allows them to pass through various obstacles and travel long distances, ideal for broadcast media.

Moderate Frequencies (Infrared, Visible Light): These frequencies involve interactions that can produce heat or stimulate electron movement, affecting how we perceive light or heat.

High-Frequency Waves (X-rays, Gamma Rays): High frequencies give these waves enough energy to break chemical bonds, which can be both beneficial (such as targeting cancer cells in therapy) and potentially harmful (risk of radiation exposure).

Applications That Depend on Frequency

Frequency influences many technologies that rely on precise energy levels:

Wireless Communication: Different frequency bands, such as FM or AM radio, allow for various methods of signal encoding and range.

Radar: Microwaves with high frequency are used in radar to detect objects by reflecting waves off surfaces.

Medical Treatments: High-frequency gamma rays are used in radiotherapy to target and destroy cancerous cells.

Frequency helps define how waves transmit information or interact with different materials, making it critical to designing applications across industries.

4. Energy: The Capacity to Interact and Transform

The energy of electromagnetic waves, defined by the equation E=hfE = hfE=hf (where EEE is energy, hhh is Planck's constant, and fff is frequency), represents the capacity of waves to cause physical or chemical changes. As frequency increases, so does energy, allowing high-energy waves to initiate chemical reactions or ionize atoms.

Types of Energy across the Spectrum

Low-Energy Waves (Radio and Microwaves): Typically do not have enough energy to ionize atoms, which makes them safer for daily use in communication devices.

Moderate-Energy Waves (Infrared and Visible Light): Capable of causing heating effects or electron excitation but generally not harmful.

High-Energy Waves (Ultraviolet, X-rays, Gamma Rays): Possess enough energy to ionize atoms and molecules, making them powerful tools in medicine and science but also potentially hazardous.

Energy level dictates both the potential use and the precautions needed when handling these waves. For instance, the high energy in X-rays allows for clear imaging of internal structures in the body, but it also requires protective shielding to minimize harmful exposure.

Energy in Applications

The applications of different types of electromagnetic waves often exploit their energy levels:

Infrared Imaging: Infrared rays detect heat signatures, which is useful for night vision and thermal cameras.

Ultraviolet Sterilization: UV-C light has enough energy to kill bacteria and viruses, making it valuable in sterilization.

Radiotherapy: The energy of gamma rays allows them to penetrate deeply and target tumors in cancer patients.

The energy carried by these waves enables various interactions with matter, whether by heating, ionizing, or passing through materials to capture information.

5. Practical Applications: Harnessing Invisible Rays Across Industries

Invisible rays have transformed industries and continue to unlock new possibilities. Below are some examples of how different types of electromagnetic waves are used in practice.

Communication and Broadcasting

Radio and Microwaves: From radio broadcasting to Wi-Fi, low-energy waves allow for global communication systems.

Television: UHF and VHF frequencies transmit video and audio signals to TVs worldwide, enabling mass communication.

Medical Imaging and Treatment

X-rays and Gamma Rays: X-ray machines provide detailed imaging for diagnosis, while gamma rays are used in cancer treatment due to their ability to target and kill cells.

Remote Sensing and Imaging

Infrared Rays: Used in satellite imaging, infrared rays reveal heat signatures of geological, environmental, and urban areas, assisting in areas such as agriculture, climate monitoring, and urban planning.

Microwaves in Radar: Microwave frequencies allow radar systems to detect the speed, position, and distance of objects, essential in aviation and meteorology.

Security and Safety

X-rays in Security Screening: Used in airport security, X-rays penetrate luggage to reveal hidden objects, improving safety.

Ultraviolet Light for Disinfection: UV-C light is used to sterilize medical equipment and public spaces, reducing pathogen spread.

6. The Future of Electromagnetic Wave Applications

As technology advances, so too will the applications of invisible rays. In healthcare, we may see increased use of targeted radiation treatments for specific types of cancers, reducing side effects and improving outcomes. In communication, research into higher frequency bands promises faster, more efficient data transmission, paving the way for technologies like 6G networks. In space exploration, radio and microwave technology will continue to expand our reach, enabling detailed study of distant galaxies and potentially providing communication for human missions to Mars.

In summary, the physics behind wavelengths, frequencies, and energy helps us to not only understand these invisible rays but also harness their unique properties. From the everyday conveniences of radio broadcasts and mobile networks to the lifesaving capabilities of medical imaging and beyond, the power of invisible rays is foundational to modern science and technology.

12. HISTORICAL DISCOVERIES

The Pioneers Who Revealed the Unseen

Throughout history, humanity has been intrigued by the concept of invisible forces and unseen energies. While ancient thinkers and early natural philosophers speculated about the forces governing the natural world, it was not until the advent of scientific inquiry and experimentation that the true nature of these invisible rays and their remarkable applications began to unfold. This chapter explores the discoveries of the pioneering scientists who revealed the unseen spectrum of rays—each expanding our understanding of physics, medicine, and the cosmos. From X-rays and radio waves to gamma rays, these breakthrough discoveries have revolutionized science, technology, and medicine, giving us tools to look deeper into the universe and within ourselves.

1. Wilhelm Röntgen and the Discovery of X-Rays

In 1895, German physicist Wilhelm Conrad Röntgen stumbled upon one of the most transformative discoveries in medical science and physics—X-rays. While experimenting with cathode rays in a darkened room, Röntgen noticed a fluorescent glow on a nearby screen. He discovered that these "X-rays" could pass through various materials and create images of objects within. The "X" signified the unknown nature of these rays at the time.

Röntgen's discovery immediately caught the attention of the scientific community and the public. The first X-ray image ever taken was of his wife's hand, showing her bones and wedding ring—a revelation that astonished scientists and laypeople alike. X-rays became a revolutionary tool in medical diagnostics, allowing doctors to view the internal structure of the body without surgery. Röntgen was awarded the first Nobel Prize in Physics in 1901 for his work, and his discovery laid the foundation for radiology, a medical field that has saved countless lives. Röntgen's contributions also spurred other researchers to explore unknown rays, leading to a cascade of discoveries in the electromagnetic spectrum.

2. Heinrich Hertz and the Confirmation of Electromagnetic Waves

German physicist Heinrich Hertz was instrumental in validating James Clerk Maxwell's theory of electromagnetic waves. Maxwell had predicted the existence of electromagnetic waves in the mid-19th century, proposing that electric and magnetic fields could propagate through space. However, Maxwell's theory remained theoretical until Hertz experimentally confirmed it in the 1880s.

Hertz designed experiments to produce and detect electromagnetic waves, demonstrating that they traveled at the speed of light and behaved like light waves. Hertz's work confirmed that light, radio waves, and other invisible rays were all forms of electromagnetic radiation. Though Hertz himself believed his discoveries had no practical use, they laid the groundwork for modern telecommunications, radio, and television. His name became synonymous with frequency measurement, with the unit "hertz" (Hz) commemorating his groundbreaking contributions.

3. Henri Becquerel and the Discovery of Radioactivity

While studying phosphorescence in uranium salts in 1896, French physicist Henri Becquerel discovered a form of radiation unlike any previously observed. Becquerel had initially hypothesized that uranium absorbed sunlight and re-emitted it as X-rays. To test this, he placed uranium salts on photographic plates covered in black paper, expecting sunlight to be necessary for the effect. However, he found that the uranium emitted radiation strong enough to expose the plates even without sunlight.

This discovery marked the beginning of our understanding of radioactivity, a phenomenon later explored by Marie and Pierre Curie. Becquerel's findings showed that some materials emit particles and energy spontaneously, fundamentally altering our understanding of atomic structure and matter itself. In 1903, Becquerel shared the Nobel Prize in Physics with the Curies for their pioneering work on radioactivity, laying the groundwork for nuclear physics and its applications in medicine and energy.

4. Marie and Pierre Curie and the Study of Radioactive Elements

Marie Curie, along with her husband Pierre, expanded upon Becquerel's work by isolating radioactive elements and studying their properties. Marie Curie coined the term "radioactivity" and discovered two new elements—polonium and radium—both of which exhibited intense radioactive properties. Her tireless research into radioactive materials provided insights into the atom's internal structure and the energy contained within.

The Curies' work had far-reaching implications, as radium became one of the first radioactive elements used in medical treatments, specifically in cancer therapy. Despite the personal risks (Marie Curie ultimately succumbed to illnesses associated with radiation exposure), their work was groundbreaking. Marie Curie was awarded two Nobel Prizes, in Physics and Chemistry, for her contributions, making her one of the most celebrated scientists in history. Their pioneering work on radioactivity ultimately paved the way for both nuclear power and medical radiation treatments.

5. Ernest Rutherford and the Understanding of Alpha, Beta, and Gamma Rays

Known as the father of nuclear physics, New Zealand-born physicist Ernest Rutherford conducted experiments that identified and characterized different types of radiation—alpha, beta, and gamma rays. While working with radioactive elements, Rutherford discovered that uranium emitted particles with distinct properties. He classified these emissions as alpha (positively charged), beta (negatively charged), and gamma rays (neutral).

Rutherford's experiments provided a deeper understanding of radioactive decay and atomic structure. His discovery of the atomic nucleus further revolutionized atomic theory. Rutherford's identification of alpha and beta particles also influenced the development of nuclear energy and medicine, with alpha particles being instrumental in cancer treatment and beta particles used in medical imaging. His work set the stage for future research into nuclear fission, which would have enormous implications for energy and warfare.

6. Paul Villard and the Discovery of Gamma Rays

In 1900, French chemist and physicist Paul Villard discovered gamma rays while studying radiation from radium. Unlike alpha and beta particles, Villard identified gamma rays as high-energy electromagnetic waves, similar to X-rays but with even greater penetrating power. Gamma rays are capable of passing through dense materials, which makes them both useful in medical treatment and challenging to shield.

The discovery of gamma rays revealed yet another part of the electromagnetic spectrum and has had profound applications in both medicine and astronomy. In medicine, gamma rays are used in cancer treatment through precise targeting to avoid healthy tissue. In astronomy, gamma-ray detection has provided insights into high-energy cosmic events, such as supernovae and black holes. Villard's work thus extended our understanding of radiation and demonstrated the existence of high-energy waves beyond visible light.

7. James Clerk Maxwell: The Theorist Behind Electromagnetic Waves

While many of the pioneers discussed thus far were experimentalists, James Clerk Maxwell's theoretical work laid the essential groundwork for understanding electromagnetic radiation. In the mid-19th century, Maxwell developed a set of equations that described the interrelation of electric and magnetic fields. These equations predicted the existence of electromagnetic waves, including those beyond visible light, such as radio waves and gamma rays.

Maxwell's theories unified electricity, magnetism, and optics into a single framework and provided the theoretical basis for all subsequent research into electromagnetic waves. His work was a keystone in physics, providing a model that would be confirmed by later experimentalists like Hertz. Maxwell's equations remain fundamental to physics, engineering, and technology, shaping everything from modern wireless communication to particle physics.

8. The Legacy and Impact of the Pioneers of Invisible Rays

The work of these pioneers—Röntgen, Hertz, Becquerel, the Curies, Rutherford, Villard, and Maxwell—opened the door to a world previously hidden from human perception. Their discoveries not only expanded our understanding of physics but also laid the foundation for revolutionary technologies that shape our everyday lives. Medical imaging, cancer treatments, wireless communication, and nuclear energy are just a few of the fields transformed by their insights into invisible rays.

Their contributions remind us that the pursuit of knowledge often leads to unexpected breakthroughs that can transform societies. As we continue to explore and harness invisible rays in fields like quantum mechanics, particle physics, and astrophysics, we build upon the legacy of these early pioneers.

9. Future Directions: Beyond the Known Spectrum

The story of invisible rays is ongoing, with modern physicists exploring realms beyond even gamma rays, such as cosmic rays and neutrinos. Advanced telescopes, particle accelerators, and detectors continue to push the boundaries of what we know about the universe, revealing new particles and forms of energy. Research into quantum mechanics, dark matter, and dark energy may lead to future discoveries that will once again redefine our understanding of the unseen forces that govern our universe.

As we stand on the shoulders of giants, the work of these historical pioneers inspires current and future scientists to continue probing the mysteries of the universe, revealing not just the invisible rays that permeate our world, but also the deeper principles that connect us to the cosmos. The discoveries of these pioneers underscore the power of curiosity, persistence, and the scientific spirit in unraveling the mysteries of the unseen.

13. RADIATION AND HEALTH

The Science of Safety and Risk Management

Radiation, a form of energy that travels through space and matter, has profoundly shaped our technological, medical, and industrial landscape. It has revolutionized fields from healthcare to energy production, yet with its immense potential comes significant responsibility. Exposure to radiation can impact human health, and as we have come to understand more about these effects, science has developed methods to assess, manage, and mitigate associated risks. This chapter explores the types of radiation, their biological effects, and the science of safety and risk management in ensuring that our interactions with radiation are as safe as possible.

1. Types of Radiation and Their Sources

Radiation exists across a broad spectrum of energy levels, from non-ionizing to ionizing radiation. These two categories differ not only in energy levels but also in the effects they can have on biological systems.

Non-Ionizing Radiation

Non-ionizing radiation includes forms of radiation such as radio waves, microwaves, infrared, and visible light. These waves have relatively low energy and are typically not harmful to humans in controlled doses. Non-ionizing radiation is omnipresent, coming from natural sources like sunlight and artificial sources such as cell phones and Wi-Fi routers. However, excessive exposure to certain types, like ultraviolet (UV) radiation, can still lead to harmful effects, such as skin damage and an increased risk of skin cancer.

Ionizing Radiation

Ionizing radiation, on the other hand, has enough energy to remove tightly bound electrons from atoms, creating ions. This category includes alpha particles, beta particles, gamma rays, and X-rays. These forms of radiation have a high potential to damage cellular structures and DNA, making them both valuable in medical applications (e.g., imaging and cancer treatment) and hazardous when uncontrolled. Sources of ionizing radiation include natural sources like radon gas, cosmic rays, and artificial sources such as nuclear reactors and medical imaging equipment.

Understanding these sources and types of radiation is essential for evaluating the health impacts associated with exposure, as each type interacts differently with biological tissues.

2. Biological Effects of Radiation Exposure

The effects of radiation on the human body depend on the type of radiation, its energy, the duration of exposure, and the tissues affected. Ionizing radiation, in particular, can have significant effects at the cellular and molecular levels.

Acute vs. Chronic Exposure

Radiation exposure can be either acute (short-term and high dose) or chronic (long-term and low dose). Acute exposure can lead to immediate health effects, including radiation sickness, burns, and even death at extremely high levels. These effects are due to the rapid damage to cells and tissues that cannot repair themselves quickly enough. Chronic exposure, by contrast, involves lower doses over an extended period. Although it may not produce immediate symptoms, prolonged exposure increases the risk of long-term effects, such as cancer and genetic mutations.

Cellular Damage and DNA Mutations

Ionizing radiation affects cells by damaging the DNA within, leading to mutations that can trigger cancer. The extent of damage depends on the type of radiation: alpha particles, while less penetrative, are highly ionizing and can cause considerable damage if inhaled or ingested. Gamma rays and X-rays, being more penetrative, can impact tissues deeper within the body. If radiation causes DNA breaks and the body cannot repair them correctly, the cells may die, function abnormally, or become cancerous.

Radiosensitivity of Different Tissues

Certain tissues are more radiosensitive, meaning they are more susceptible to radiation damage. For example, bone marrow, reproductive organs, and gastrointestinal tissues are highly radiosensitive. Children and younger individuals are generally more vulnerable to radiation because their cells are dividing more rapidly. Therefore, medical imaging involving radiation is often used more sparingly in pediatric care.

Understanding these biological effects is foundational to developing safety protocols and establishing permissible exposure limits to minimize health risks.

3. Measuring Radiation Exposure: Units and Dosimetry

Radiation exposure is measured using specific units and devices that help quantify the dose absorbed by human tissues. The primary units include:

Gray (Gy): Measures the absorbed dose of radiation by matter, typically tissue.

Sievert (Sv): Adjusts the absorbed dose to reflect the biological impact, factoring in the type of radiation and the tissues affected.

The use of dosimeters—devices that measure cumulative radiation exposure—is common in industries where workers are regularly exposed to radiation, such as healthcare, nuclear power, and aerospace.

Personal dosimetry, area monitoring, and environmental radiation monitoring are part of a broader system to ensure that exposure levels remain within safe limits.

4. Radiation Safety Standards and Guidelines

Radiation protection standards are developed by international organizations, national agencies, and regulatory bodies based on extensive research. Organizations such as the International Commission on Radiological

Protection (ICRP), the National Council on Radiation Protection and Measurements (NCRP), and the United Nations Scientific Committee on the Effects of Atomic Radiation (UNSCEAR) provide guidelines to minimize risk.

These guidelines generally follow three fundamental principles:

Justification: Any activity involving radiation exposure must have a justified benefit that outweighs potential risks.

Optimization (ALARA Principle): Exposure should be kept "As Low As Reasonably Achievable," balancing safety with practicality.

Dose Limitation: There are established dose limits for occupational exposure, public exposure, and medical exposure, designed to minimize health risks without hindering beneficial applications.

Each country adapts these recommendations to its regulations, establishing permissible exposure levels for different groups (e.g., workers, the general public, and medical patients).

5. Managing Occupational Exposure: Protocols and Protections

For those working in environments with potential radiation exposure, strict safety protocols are essential. This includes individuals in the medical field (e.g., radiologists, technicians), nuclear industry workers, and researchers in radiation-related fields. Safety measures in these workplaces include:

Protective Equipment: Lead aprons, shields, and barriers help reduce exposure to ionizing radiation, especially in healthcare settings.

Time, Distance, and Shielding: These are the three core principles in radiation protection. Minimizing time spent near radiation sources, maximizing distance from sources, and using adequate shielding (such as lead walls) significantly reduce exposure.

Regular Training and Education: Workers receive training on radiation safety, risks, and best practices for minimizing exposure.

Routine Monitoring and Health Check-ups: Regular dosimetry checks and health screenings are conducted to track cumulative radiation exposure, ensuring it stays within safe levels.

In the case of nuclear power plants, extensive safety systems and protocols are in place to prevent radiation leakage, including containment structures, automatic shutdown systems, and emergency response plans.

6. Medical Exposure and Patient Safety

Radiation is invaluable in medicine, but it must be carefully controlled to protect patients from unnecessary exposure. The primary applications in healthcare include diagnostic imaging (e.g., X-rays, CT scans) and radiation therapy for cancer treatment.

Diagnostic Imaging

X-rays, CT scans, and fluoroscopy allow physicians to see inside the body, but overuse can lead to cumulative exposure risks. Radiologists use the ALARA principle to limit dose exposure during imaging. Protective measures, such as lead shields, are used to cover sensitive areas of the body, and alternative imaging methods (e.g., ultrasound, MRI) are considered when appropriate.

Radiation Therapy

In cancer treatment, high doses of radiation are targeted precisely to destroy cancerous cells while minimizing exposure to surrounding healthy tissue. Advances in radiation therapy, including techniques like intensity-modulated radiation therapy (IMRT) and proton therapy, allow for greater accuracy and fewer side effects. Radiotherapy specialists and medical physicists play crucial roles in planning and delivering safe treatments.

7. Public Safety and Environmental Protection

Radiation also exists in our environment from natural and artificial sources. Radon gas, cosmic rays, and fallout from nuclear testing all contribute to background radiation. Managing public exposure involves monitoring these levels and reducing risks where possible.

For instance, building codes and ventilation systems are now designed to minimize radon exposure in homes. The aftermath of nuclear accidents, such as the Chernobyl and Fukushima disasters, demonstrated the importance of having emergency response plans and decontamination procedures to protect the public from radioactive contamination.

8. Advancements in Radiation Risk Management

With ongoing research, new technologies and methods are constantly being developed to improve radiation safety and risk management:

Digital Imaging: In medical diagnostics, digital X-rays and advanced CT scans require lower doses than traditional methods, minimizing patient exposure.

Wearable Dosimeters: Modern dosimeters are smaller and more accurate, allowing individuals to monitor their exposure in real-time.

Radiation Protection Software: Hospitals and nuclear facilities utilize software to track and predict radiation exposure, optimizing safety for both workers and patients.

Biomarkers for Radiation Exposure: Research is underway to develop biomarkers that indicate radiation exposure levels, which could help assess risk more accurately in the aftermath of an incident.

9. Future Challenges and Ethical Considerations

As we continue to harness radiation for new technologies, we face ethical challenges regarding its applications. The expansion of nuclear energy and the development of advanced medical treatments must be balanced with concerns about environmental impact, radiation waste, and equitable access to safe treatments.

Additionally, as radiation-related technology advances, society will need to continuously update regulatory frameworks and safety standards to address emerging risks and ensure the responsible use of radiation.

Conclusion

Radiation's dual role—as both a powerful tool and a potential hazard—necessitates an ongoing commitment to safety and risk management. Through diligent adherence to safety principles, continuous advancements in technology, and a profound understanding of the biological impacts of radiation, we can leverage the benefits of radiation while safeguarding human health and the environment. The science of radiation safety is one of constant vigilance and adaptation, underscoring the importance of responsibility in our pursuit of progress.

14. RAYS IN MEDICINE

Transforming Diagnostics and Treatments

The field of medicine has been revolutionized by the discovery and utilization of various forms of rays, from X-rays to gamma rays, which have fundamentally changed the way we diagnose, treat, and manage diseases. These invisible forms of energy have enabled physicians to peer inside the human body, uncovering details that were once hidden from view. Over the past century, medical rays have transformed from rudimentary diagnostic tools into sophisticated technologies that save countless lives each year. In this chapter, we explore the history, science, and groundbreaking applications of rays in medicine, as well as the future possibilities and challenges they present.

A Brief History of Medical Rays

The journey of using rays in medicine began with the accidental discovery of X-rays by Wilhelm Conrad Roentgen in 1895. Roentgen noticed that a mysterious form of radiation could pass through solid objects and produce images on photographic plates. He famously captured the first X-ray image of his wife's hand, revealing the bones within. This discovery earned him the first Nobel Prize in Physics in 1901 and marked the birth of medical imaging.

Following the discovery of X-rays, Henri Becquerel and Marie Curie discovered radioactivity, leading to the identification of alpha, beta, and gamma rays. These breakthroughs laid the foundation for the use of rays in both diagnostics and treatment. The subsequent decades saw the development of technologies like computed tomography (CT), magnetic resonance imaging (MRI), and positron emission tomography (PET) scans, which expanded the capabilities of medical imaging far beyond what was previously imaginable.

The Science behind Medical Rays

Rays used in medicine can be classified into two main categories: ionizing and non-ionizing radiation. Ionizing radiation, which includes X-rays and gamma rays, has enough energy to remove tightly bound electrons from atoms, creating ions. This ability to ionize atoms makes these rays highly effective for medical imaging and cancer treatments, but it also means they must be used with caution due to potential health risks.

Non-ionizing radiation, on the other hand, includes ultraviolet (UV) rays, infrared (IR) rays, and radio waves. These rays do not carry enough energy to ionize atoms and are generally considered safer for diagnostic applications like MRI and ultrasound imaging. Each type of ray interacts with tissues in unique ways, allowing for diverse medical applications.

Diagnostic Imaging: Seeing the Invisible

The most widespread use of rays in medicine is in diagnostic imaging. These techniques allow healthcare professionals to visualize the internal structures of the body, aiding in the diagnosis and monitoring of diseases. Let's explore some of the most common imaging modalities:

1. X-Ray Imaging

X-ray imaging is one of the oldest and most widely used forms of diagnostic imaging. By passing X-rays through the body, it produces images of internal structures based on the varying absorption rates of different tissues. Bones, which are dense, appear white on X-ray images, while softer tissues, such as muscles and organs, appear in shades of gray.

Applications:

Fracture Detection: X-rays are invaluable for identifying bone fractures and dislocations.

Chest Radiography: Used to diagnose conditions like pneumonia, tuberculosis, and lung cancer.

Dental Imaging: Helps dentists identify cavities, root infections, and jawbone issues.

Advancements: Modern digital X-ray systems have improved image quality, reduced radiation exposure, and enabled the storage of images in electronic medical records for easy sharing and analysis.

2. Computed Tomography (CT) Scans

CT scans, also known as CAT scans, use a series of X-ray images taken from different angles to produce cross-sectional images of the body. These images are then processed by a computer to create detailed 3D representations of internal organs and tissues.

Applications:

Cancer Detection: CT scans are highly effective in identifying tumors, assessing their size, and monitoring their progression.

Trauma Care: Essential in emergency settings for diagnosing internal injuries, brain hemorrhages, and abdominal trauma.

Cardiac Imaging: Used to visualize blood vessels, detect blockages, and assess heart function.

Advancements: The development of low-dose CT scans has minimized radiation exposure, making it safer for routine screening, such as lung cancer screening in high-risk individuals.

3. Magnetic Resonance Imaging (MRI)

MRI uses powerful magnets and radio waves to generate detailed images of soft tissues, including the brain, spinal cord, and joints. Unlike X-rays and CT scans, MRI does not use ionizing radiation, making it safer for certain applications.

Applications:

Neurology: MRI is the gold standard for diagnosing conditions like multiple sclerosis, brain tumors, and spinal cord injuries.

Orthopedics: Useful for visualizing ligaments, tendons, and cartilage in cases of sports injuries or arthritis.

Cardiology: Cardiac MRI can assess heart muscle function, detect congenital heart defects, and evaluate the severity of heart disease.

Advancements: Functional MRI (fMRI) is a cutting-edge technique that measures brain activity by detecting changes in blood flow, helping researchers understand brain function and map neurological diseases.

4. Positron Emission Tomography (PET) Scans

PET scans involve injecting a small amount of radioactive tracer into the body to visualize metabolic activity. The tracer emits positrons, which collide with electrons to produce gamma rays that are detected by the scanner.

Applications:

Oncology: PET scans are highly sensitive for detecting cancer metastasis, assessing tumor response to treatment, and planning radiotherapy.

Neurology: Used to diagnose Alzheimer's disease, epilepsy, and Parkinson's disease by observing metabolic changes in the brain.

Cardiology: Helps evaluate blood flow to the heart muscle, identify areas of ischemia, and assess myocardial viability.

Advancements: Combining PET with CT or MRI (PET-CT or PET-MRI) provides both functional and anatomical information, improving diagnostic accuracy.

Therapeutic Uses of Rays: Beyond Diagnostics

While diagnostic imaging is perhaps the most familiar application of rays in medicine, their therapeutic uses are equally transformative. Let's explore how rays are used to treat a variety of conditions.

1. Radiation Therapy for Cancer

Radiation therapy is a cornerstone of cancer treatment, used to target and destroy cancer cells while sparing surrounding healthy tissue. It works by damaging the DNA within cancer cells, preventing them from replicating and growing.

Types of Radiation Therapy:

External Beam Radiation Therapy (EBRT): Uses high-energy X-rays or proton beams directed at the tumor from outside the body. Techniques like intensity-modulated radiation therapy (IMRT) allow for precise targeting, minimizing damage to nearby tissues.

Brachytherapy: Involves placing radioactive sources directly into or near the tumor. Commonly used for prostate, cervical, and breast cancers, brachytherapy delivers high doses of radiation with minimal side effects.

Stereotactic Radiosurgery (SRS): A highly precise form of radiation therapy used for treating brain tumors and other small lesions with minimal impact on surrounding tissues.

Advancements: The advent of proton therapy and carbon ion therapy, which use charged particles instead of photons, allows for even greater precision in targeting tumors while reducing collateral damage.

2. Radioactive Isotopes in Medicine

Radioactive isotopes, also known as radionuclides, have a wide range of medical applications. One of the most well-known is the use of Iodine-131 to treat thyroid disorders, including hyperthyroidism and thyroid cancer. Radionuclides are also used in diagnostic imaging (e.g., technetium-99m in nuclear medicine scans) and pain relief for bone metastases.

3. Laser and Ultraviolet (UV) Rays in Dermatology

Lasers and UV rays are widely used in dermatology for both therapeutic and cosmetic procedures. For instance, UV light therapy (phototherapy) is used to treat skin conditions like psoriasis, eczema, and vitiligo. Laser treatments are popular for removing skin lesions, reducing wrinkles, and even treating certain types of cancer.

Safety and Ethical Considerations

The use of rays in medicine, particularly ionizing radiation, presents certain risks, including the potential for radiation-induced cancer. Therefore, it is essential to balance the benefits of medical rays with their risks, following strict guidelines to minimize exposure.

1. Radiation Protection Principles

The principle of ALARA (As Low As Reasonably Achievable) guides radiation safety practices, aiming to minimize exposure to patients and healthcare workers. Protective measures include:

Shielding: Using lead aprons, thyroid collars, and barriers to block radiation.

Dose Optimization: Adjusting radiation doses to the minimum necessary for effective imaging or treatment.

Regular Monitoring: Monitoring radiation exposure levels for healthcare professionals to ensure they stay within safe limits.

2. Ethical Issues in Radiation Use

Ethical considerations in the use of medical rays include ensuring informed consent, particularly when using high-dose procedures like CT scans or radiation therapy. Patients should be fully informed of the benefits and potential risks of radiation exposure.

Moreover, the accessibility of advanced imaging and radiation therapy is a growing concern, particularly in low-income regions where these technologies are less available. Efforts to improve global access to life-saving radiological technologies are essential for advancing healthcare equity.

The Future of Rays in Medicine

The future of rays in medicine is filled with exciting possibilities, driven by ongoing research and technological advancements. Emerging areas include:

Artificial Intelligence (AI) in Imaging: AI algorithms can enhance image analysis, enabling faster and more accurate diagnoses.

Molecular Imaging: Techniques like PET-MRI are becoming more sophisticated, allowing for earlier detection of diseases at the molecular level.

Personalized Radiotherapy: Advances in genomics and radiobiology are paving the way for personalized radiation therapy, tailored to an individual's unique tumor biology.

From the first X-ray images taken over a century ago to the cutting-edge radiation therapies of today, the use of rays in medicine has transformed healthcare in profound ways. These invisible forms of energy have opened new windows into the human body, enabling earlier diagnoses, more precise treatments, and ultimately, better patient outcomes.

As we continue to harness the power of rays, we must also remain vigilant in managing their risks and ensuring that these life-saving technologies are accessible to all. The future of medicine will undoubtedly see rays playing an even greater role in diagnostics, therapy, and beyond, as we continue to explore the unseen dimensions of the human body in our quest for better health.

15. INDUSTRIAL APPLICATIONS

From Material Testing to Quality Control

In the industrial sector, the use of invisible rays—from X-rays to gamma rays—has revolutionized various processes, including material testing, quality control, and manufacturing. These applications extend far beyond the medical field, offering powerful tools that ensure product safety, enhance quality, and optimize efficiency. By leveraging the unique properties of different types of rays, industries are able to achieve non-destructive testing, accurate measurements, and precise inspections. In this chapter, we explore the historical development, scientific principles, and diverse industrial applications of rays, as well as their future potential.

A Brief History of Rays in Industry

The journey of utilizing rays in industrial applications began shortly after Wilhelm Conrad Roentgen discovered X-rays in 1895. While the medical community quickly recognized the diagnostic potential of X-rays, the industrial sector also saw the opportunity to apply this technology to inspect materials and components without damaging them. By the early 20th century, X-ray imaging had become a vital tool for detecting internal flaws in metal castings, welds, and other critical components.

The discovery of radioactivity by Henri Becquerel, followed by the pioneering research of Marie and Pierre Curie, introduced gamma rays to the industrial world. Due to their high energy and penetrating power, gamma rays became essential for inspecting dense materials and structures. This laid the foundation for the field of non-destructive testing (NDT), which has since expanded to include various other types of rays, such as infrared, ultraviolet, and microwaves.

The Science behind Industrial Rays

To understand how rays are utilized in industrial applications, it's important to grasp their fundamental properties. Rays can be classified into two main categories: ionizing and non-ionizing radiation.

1. Ionizing Radiation

Ionizing radiation includes X-rays and gamma rays, which have enough energy to ionize atoms and molecules. This property makes them particularly useful for penetrating dense materials and revealing internal structures. However, due to their ionizing nature, they must be handled with care to avoid potential health hazards.

2. Non-Ionizing Radiation

Non-ionizing radiation encompasses rays such as infrared, ultraviolet, and microwaves. These rays do not have sufficient energy to ionize atoms, making them safer for certain applications. They are commonly used for surface inspections, thermal imaging, and communication technologies in industrial settings.

Each type of ray interacts with materials in unique ways, enabling a wide range of applications across various industries.

Key Industrial Applications of Rays

The use of rays in industrial applications spans numerous sectors, from aerospace and automotive manufacturing to oil and gas exploration. Below, we delve into some of the most significant uses of rays in industry.

1. Non-Destructive Testing (NDT)

Non-destructive testing (NDT) is one of the most crucial applications of rays in industry. NDT methods allow for the examination of materials and components without causing any damage, making them ideal for quality control, safety assessments, and maintenance checks.

X-Ray and Gamma Ray Radiography

Material Inspection: X-ray and gamma ray radiography are widely used to detect internal flaws such as cracks, voids, and inclusions in metal castings, welds, and pipelines. This technique is especially important in the aerospace, automotive, and construction industries, where structural integrity is critical.

Weld Inspection: Ensuring the quality of welds is essential for safety in structures like bridges, ships, and oil rigs. Radiographic inspection helps identify defects such as incomplete fusion, porosity, and slag inclusions.

Pressure Vessel Testing: Radiography is used to inspect pressure vessels, boilers, and storage tanks for corrosion, cracks, and other defects that could lead to catastrophic failures.

Ultrasonic Testing

While not a ray in the traditional sense, ultrasonic waves are another form of non-destructive testing that complements radiography. High-frequency sound waves are used to detect flaws in materials, measure thickness, and assess bond quality. Ultrasonic testing is particularly effective for detecting defects in composite materials, which are increasingly used in aerospace and automotive manufacturing.

2. Industrial Radiography in Oil and Gas

The oil and gas industry relies heavily on rays for the inspection and maintenance of pipelines, refineries, and drilling equipment. Gamma ray radiography is especially useful for inspecting the integrity of welds in pipelines, which are often buried underground or submerged in water. This non-invasive technique helps prevent leaks and ruptures that could lead to environmental disasters.

Pipeline Corrosion Monitoring

Corrosion is a major concern in the oil and gas industry. X-ray and gamma ray techniques are used to monitor the thickness of pipeline walls, detect corrosion under insulation, and assess the overall condition of pipelines. These inspections are critical for ensuring the safe transport of oil and gas over long distances.

3. X-Ray Fluorescence (XRF) Spectroscopy

X-ray fluorescence (XRF) spectroscopy is a powerful analytical technique used to determine the elemental composition of materials. When a material is exposed to X-rays, it emits secondary (fluorescent) X-rays that are characteristic of its elemental composition. XRF is widely used in various industries for:

Mining and Geology: Identifying and quantifying the presence of valuable minerals and ores in rock samples.

Metal Recycling: Sorting and analyzing scrap metals to determine their alloy composition for recycling purposes.

Environmental Testing: Detecting heavy metals in soil, water, and air samples for environmental monitoring and compliance with regulations.

4. Infrared Thermography

Infrared (IR) thermography is a non-contact technique that uses infrared radiation to detect temperature variations on the surface of objects. This technology is extensively used for:

Predictive Maintenance: Identifying overheating components in electrical systems, mechanical equipment, and industrial machinery before they fail.

Building Inspections: Detecting heat leaks, moisture intrusion, and insulation defects in buildings to improve energy efficiency.

Quality Control: Inspecting the thermal properties of products during manufacturing, such as detecting defects in plastic moldings, electronic circuits, and glass products.

5. Ultraviolet (UV) Inspection

Ultraviolet (UV) rays are used in industrial applications to detect surface defects, contamination, and leaks. UV inspection is particularly effective for:

Detecting Cracks and Surface Flaws: UV light reveals cracks and other surface defects in ceramics, glass, and metal components by causing fluorescent dyes to glow.

Leak Detection: UV dyes are added to fluids in hydraulic systems, refrigeration units, and pipelines. Leaks are then identified by shining UV light, which makes the dye visible.

Quality Assurance in Manufacturing: UV inspection ensures that coatings, adhesives, and sealants have been applied correctly by revealing any gaps or inconsistencies.

6. Microwaves in Industrial Processing

Microwaves are widely used in industrial applications for heating, drying, and material processing. The advantages of microwave processing include faster heating times, improved energy efficiency, and selective heating of specific materials.

Food Processing: Microwaves are used for pasteurization, sterilization, and drying of food products, extending their shelf life without compromising nutritional quality.

Material Synthesis: Microwaves enable the rapid synthesis of advanced materials, including ceramics, polymers, and nanomaterials, which are used in electronics, aerospace, and medical devices.

Moisture Measurement: Microwave sensors are used to measure the moisture content of materials in real-time, which is essential for quality control in industries such as paper, textiles, and pharmaceuticals.

Quality Control and Automation

The integration of rays into quality control processes has led to significant improvements in product consistency, safety, and efficiency. Automated systems equipped with ray-based sensors and imaging technologies can perform continuous inspections on production lines, reducing the likelihood of defects and enhancing overall quality.

1. Real-Time X-Ray Inspection Systems

Real-time X-ray inspection systems are used in industries like automotive, electronics, and aerospace to inspect components as they are being produced. These systems can detect hidden defects, such as air pockets in castings or solder joint failures in electronic circuits, with high accuracy.

2. Automated Infrared Inspection

Infrared cameras integrated with artificial intelligence (AI) algorithms are used for automated inspection of products, such as detecting anomalies in the thermal profiles of electronic devices, identifying hotspots in photovoltaic panels, and assessing the quality of welds in automotive manufacturing.

Safety and Environmental Considerations

While the industrial use of rays offers numerous benefits, it also presents safety and environmental challenges that must be addressed.

1. Radiation Safety

Industries that use ionizing radiation, such as X-rays and gamma rays, must adhere to strict safety protocols to protect workers from exposure. This includes the use of lead shielding, protective gear, and radiation monitoring devices.

2. Environmental Impact

The disposal of radioactive materials used in industrial applications, such as gamma ray sources in radiography, requires careful handling to prevent environmental contamination. Regulatory agencies set guidelines for the safe disposal and recycling of radioactive waste.

Future Trends and Innovations

The future of industrial applications of rays is poised for significant advancements, driven by innovations in AI, machine learning, and nanotechnology. Emerging trends include:

AI-Driven Image Analysis: AI algorithms can enhance the interpretation of X-ray, infrared, and ultrasound images, enabling faster and more accurate defect detection.

Nanotechnology in Radiation Detection: Advances in nanomaterials are leading to the development of more sensitive and efficient radiation detectors, which can improve safety monitoring and quality control.

Portable NDT Devices: The miniaturization of NDT equipment is enabling on-site inspections in remote locations, reducing downtime and improving maintenance efficiency.

From ensuring the safety of bridges and pipelines to enhancing the quality of consumer products, the use of rays in industrial applications has become indispensable. These invisible forms of energy offer unparalleled insights into the integrity and composition of materials, driving innovations that improve efficiency, safety, and sustainability. As industries continue to embrace cutting-edge technologies, the role of rays in material testing, quality control, and process optimization will only grow, shaping the future of manufacturing and beyond.

Through ongoing research and development, the industrial sector will continue to harness the power of rays, pushing the boundaries of what is possible and paving the way for new applications that enhance our world.

16. RAYS IN SCIENTIFIC RESEARCH

Unveiling the Secrets of the Universe

In the vast expanse of the cosmos and the microscopic realms of matter, invisible rays have played an extraordinary role in advancing scientific knowledge. From X-rays and gamma rays to ultraviolet and infrared, these forms of electromagnetic radiation have become indispensable tools in the quest to understand the universe. They allow scientists to peer into the deepest regions of space, investigate the fundamental structures of matter, and explore phenomena beyond the capabilities of visible light. This chapter delves into how different rays are harnessed in scientific research, revealing insights that were once beyond the reach of human understanding.

A Historical Perspective: The Dawn of Invisible Rays in Science

The story of rays in scientific research began in the late 19th and early 20th centuries. The discovery of X-rays by Wilhelm Conrad Roentgen in 1895 marked a turning point in the exploration of invisible rays. Roentgen's work not only transformed medical diagnostics but also inspired scientists to explore the potential of other forms of electromagnetic radiation. This led to groundbreaking discoveries, including radioactivity by Henri Becquerel and the identification of ultraviolet and infrared rays, which expanded the spectrum of scientific inquiry.

The early 20th century also witnessed the birth of quantum mechanics, as scientists like Max Planck, Niels Bohr, and Albert Einstein began to understand the particle-like properties of electromagnetic waves. This understanding laid the foundation for using rays in scientific research, from probing atomic structures to exploring the depths of space.

Understanding the Electromagnetic Spectrum

To appreciate the diverse applications of rays in scientific research, it is essential to understand their place in the electromagnetic spectrum. The electromagnetic spectrum encompasses all forms of electromagnetic radiation, arranged by wavelength and frequency. At one end of the spectrum, we have long-wavelength, low-frequency radio waves, while at the other, we find short-wavelength, high-frequency gamma rays. In between lie microwaves, infrared, visible light, ultraviolet, and X-rays.

Each type of ray interacts with matter in unique ways, allowing scientists to tailor their use to specific research objectives. For instance, radio waves are ideal for studying distant galaxies, while X-rays and gamma rays are suited for examining the atomic and subatomic levels.

Rays in Astrophysics: Exploring the Cosmos

One of the most significant scientific applications of rays is in the field of astrophysics. Astronomers and astrophysicists use various types of electromagnetic radiation to study celestial objects and cosmic phenomena. These rays provide information that is otherwise invisible to optical telescopes, revealing the hidden aspects of the universe.

1. Radio Waves: Mapping the Universe

Radio waves have the longest wavelengths in the electromagnetic spectrum, ranging from a few millimeters to several kilometers. Despite their long wavelengths, radio waves are incredibly useful in exploring the universe.

Radio Telescopes: These instruments detect radio waves emitted by celestial bodies like stars, galaxies, and even planets. The iconic Arecibo Observatory (before its collapse) and the Very Large Array (VLA) in New Mexico have been instrumental in mapping the universe, discovering pulsars, and studying the cosmic microwave background radiation.

Pulsars and Quasars: Radio waves have allowed scientists to discover pulsars—rapidly spinning neutron stars that emit beams of radio waves—

and quasars, which are among the most energetic and distant objects in the universe.

SETI (Search for Extraterrestrial Intelligence): Radio waves are also used in the search for extraterrestrial life, as scientists scan the skies for potential signals from intelligent civilizations.

2. Infrared Rays: Seeing Through the Dust

Infrared radiation has wavelengths longer than visible light but shorter than microwaves. This part of the spectrum is crucial for observing objects that are too cool to emit visible light or are obscured by cosmic dust.

Infrared Telescopes: Space-based observatories like the Hubble Space Telescope (infrared capabilities) and the James Webb Space Telescope (JWST) use infrared rays to peer through clouds of dust and gas that obscure the view in visible light. This has enabled the study of star formation, planetary systems, and the early stages of galaxy formation.

Exoplanet Detection: Infrared observations are essential for detecting exoplanets—planets orbiting stars outside our solar system. By observing the infrared light emitted or absorbed by these planets, scientists can learn about their atmospheres, compositions, and potential habitability.

Thermal Imaging: Infrared rays are also used in thermal imaging to study the heat emitted by celestial objects, providing insights into their temperature and energy output.

3. X-Rays: Investigating High-Energy Phenomena

X-rays have much shorter wavelengths and higher energy than visible light, making them ideal for studying high-energy processes in the universe.

X-Ray Astronomy: X-ray telescopes like NASA's Chandra X-ray Observatory and the European Space Agency's XMM-Newton Observatory have opened a new window into the universe. They are used to study the remnants of supernovae, black holes, neutron stars, and the hot gas found in galaxy clusters.

Black Holes and Neutron Stars: X-rays are crucial for detecting the presence of black holes and neutron stars, as these objects emit intense X-ray radiation when matter is accreted onto their surfaces.

Solar Studies: X-ray observations of the Sun reveal details about solar flares, coronal mass ejections, and other high-energy phenomena that affect space weather and, ultimately, life on Earth.

4. Gamma Rays: Unveiling the Most Violent Events

Gamma rays are the most energetic form of electromagnetic radiation and are produced by the most extreme events in the universe.

Gamma-Ray Bursts: These bursts are among the most energetic events observed in the cosmos, often associated with the collapse of massive stars or the merging of neutron stars. They release as much energy in a few seconds as the Sun will emit over its entire 10-billion-year lifespan.

Fermi Gamma-ray Space Telescope: NASA's Fermi telescope has been instrumental in studying gamma-ray bursts, active galactic nuclei, and other cosmic phenomena that produce gamma rays.

Dark Matter Research: Gamma rays are also used in the search for dark matter, as scientists look for gamma-ray signatures that could indicate the presence of this elusive substance.

Rays in Particle Physics: Probing the Building Blocks of Matter

Beyond astronomy, rays play a vital role in particle physics, where scientists seek to understand the fundamental particles and forces that make up the universe.

1. X-Ray Crystallography

X-ray crystallography is a technique that has revolutionized fields like chemistry, biology, and materials science. It involves directing X-rays at a crystal and analyzing the resulting diffraction pattern to determine the atomic structure of the material.

Protein Structure: X-ray crystallography has been instrumental in determining the structures of complex biological molecules like DNA, proteins, and enzymes. Understanding these structures is crucial for drug design and the development of new therapies.

Material Science: The technique is also used to study the arrangement of atoms in various materials, leading to innovations in fields like nanotechnology and metallurgy.

2. Particle Accelerators and Synchrotron Radiation

Particle accelerators, like the Large Hadron Collider (LHC) at CERN, use high-energy rays to collide particles at nearly the speed of light. These collisions produce a range of rays, including gamma rays, which help scientists explore the properties of subatomic particles.

Discovering New Particles: The LHC has been instrumental in discovering particles like the Higgs boson, which helps explain why other particles have mass.

Synchrotron Radiation: Synchrotron facilities use the radiation produced by accelerating electrons to study the properties of materials, biological samples, and chemical reactions with high precision.

3. Neutron and Gamma Ray Spectroscopy

Neutron and gamma ray spectroscopy are powerful tools for analyzing the composition of materials and understanding nuclear processes.

Nuclear Research: These techniques are used to study nuclear reactions, radioactive decay, and the behavior of materials under extreme conditions, such as those found in nuclear reactors and during space missions.

Planetary Exploration: Neutron and gamma ray spectroscopy have been used in space missions, such as NASA's Mars rovers, to analyze the composition of planetary surfaces and search for signs of water and other elements.

The Future of Rays in Scientific Research

As technology advances, the use of rays in scientific research continues to expand. Here are some emerging trends and future directions:

Quantum X-Ray Imaging: Researchers are developing quantum-enhanced X-ray imaging techniques that could provide higher resolution and contrast, opening new possibilities in both scientific research and medical diagnostics.

Advanced Space Telescopes: Future space telescopes, like the European Space Agency's Athena X-ray Observatory, aim to study the high-energy universe with unprecedented detail, potentially unlocking new insights into dark matter and dark energy.

Neutrino Astronomy: While not a ray in the traditional sense, neutrinos are nearly massless particles that rarely interact with matter. New detectors, like the IceCube Neutrino Observatory, aim to use these particles to study cosmic events that are invisible to other forms of radiation.

The use of rays in scientific research has fundamentally transformed our understanding of the universe, from the largest structures in the cosmos to the smallest particles of matter. By harnessing the invisible, scientists have unlocked secrets that were once beyond our reach, revealing a universe that is more complex, dynamic, and awe-inspiring than we ever imagined.

As technology continues to advance, the potential applications of rays in scientific research are boundless. From exploring the farthest reaches of space to probing the deepest mysteries of matter, the science of rays will continue to illuminate the unknown, guiding humanity toward new discoveries and innovations.

Through the lens of invisible rays, we not only see the universe in a new light but also gain a deeper appreciation for the fundamental forces that shape our world. The journey of scientific exploration is far from over, and the future holds the promise of even more remarkable discoveries, driven by the power of rays.

17. COMMUNICATION BREAKTHROUGHS

The Role of Radio and Microwaves in Modern Society

From the discovery of radio waves in the late 19th century to the development of microwave technology in the 20th century, the world has witnessed a series of communication breakthroughs that have fundamentally transformed modern society. These advancements have not only reshaped how we communicate but have also driven technological innovation across a range of industries. In this chapter of "Harnessing the Invisible: The Science and Applications of Rays from Alpha to X," we explore the profound impact of radio and microwaves on communication, their underlying science, and their diverse applications in today's interconnected world.

The Birth of Radio: A Revolution in Communication

The Discovery of Radio Waves

The journey of radio technology began with the theoretical groundwork laid by James Clerk Maxwell, who, in the 1860s, developed the equations that described electromagnetic waves. His work suggested that electromagnetic waves could travel through space at the speed of light. Building on Maxwell's theories, Heinrich Hertz, in 1887, experimentally demonstrated the existence of radio waves, proving that they could be generated and detected.

Guglielmo Marconi and the Dawn of Wireless Communication

The practical applications of radio waves were pioneered by Guglielmo Marconi, an Italian inventor who is widely regarded as the father of radio.

In 1895, Marconi succeeded in transmitting wireless signals over a distance of more than a mile. By 1901, he had achieved the first transatlantic radio transmission, bridging a distance of over 2,000 miles between England and Newfoundland. This breakthrough demonstrated the potential of radio waves for long-distance communication, sparking a global interest in the technology.

The Golden Age of Radio Broadcasting

The early 20th century saw the rise of radio broadcasting as a mass communication medium. By the 1920s, radio had become a household staple, with families gathering around to listen to news, music, and entertainment programs. The golden age of radio not only revolutionized the entertainment industry but also played a crucial role in shaping public opinion during major historical events like World War II.

The Science of Radio Waves

Understanding Radio Waves

Radio waves are a type of electromagnetic radiation with wavelengths ranging from about one millimeter to 100 kilometers, corresponding to frequencies from 3 kHz to 300 GHz. Unlike sound waves, which require a medium like air or water to travel through, radio waves can propagate through the vacuum of space, making them ideal for wireless communication.

How Radio Waves Transmit Information

The fundamental principle behind radio communication is the modulation of radio waves to encode information. There are two main types of modulation:

Amplitude Modulation (AM): In AM radio, the strength (amplitude) of the carrier wave is varied in proportion to the signal being transmitted. This method was widely used in early radio broadcasting.

Frequency Modulation (FM): In FM radio, the frequency of the carrier wave is varied according to the information signal. FM provides better sound quality and resistance to interference compared to AM, making it popular for music broadcasting.

The Role of Antennas

Antennas play a crucial role in the transmission and reception of radio waves. They convert electrical signals into radio waves for broadcasting and convert incoming radio waves back into electrical signals for reception. The design and placement of antennas significantly affect the range and quality of communication.

The Impact of Radio on Modern Society

Radio in Emergency Communication

One of the most significant contributions of radio technology is its role in emergency communication. Radio remains a reliable medium for transmitting emergency alerts, weather updates, and disaster warnings. During natural disasters like hurricanes and earthquakes, radio broadcasts can reach affected areas where internet and cellular networks may be compromised.

Radio in Navigation: The Birth of GPS

Radio waves are also the foundation of navigation systems. The development of the Global Positioning System (GPS) relies on a network of satellites that transmit radio signals to receivers on Earth. By measuring the time it takes for the signals to travel from the satellites to the receiver, GPS devices can calculate precise locations. This technology has transformed navigation for aviation, maritime, and personal travel, making it an integral part of modern life.

Radio in Scientific Exploration: The Search for Extraterrestrial Life

Radio waves have played a crucial role in the field of astronomy and the search for extraterrestrial intelligence (SETI). Radio telescopes, like the iconic Arecibo Observatory, have been used to detect cosmic phenomena such as pulsars and quasars. These telescopes also scan the skies for potential signals from intelligent civilizations beyond Earth.

The Emergence of Microwave Technology

Understanding Microwaves

Microwaves are a subset of radio waves with shorter wavelengths, typically ranging from one millimeter to one meter (frequencies between 300 MHz and 300 GHz). Due to their shorter wavelengths, microwaves can carry more data than longer-wavelength radio waves, making them ideal for high-bandwidth communication.

The Discovery and Development of Microwaves

The practical use of microwaves began during World War II with the development of radar (Radio Detection and Ranging). Radar technology uses microwave signals to detect objects' distance, speed, and direction, making it a vital tool for military applications. After the war, the focus shifted to civilian uses, leading to the development of microwave ovens, satellite communication, and wireless networks.

Microwave Communication: Connecting the World

Satellite Communication

Microwave technology has been instrumental in the development of satellite communication, which has transformed global connectivity. Communication satellites orbiting the Earth use microwave signals to relay information between ground stations. This technology enables long-distance phone calls, television broadcasts, and internet services across the globe.

The Birth of the Communication Satellite: The launch of the first communication satellite, Telstar, in 1962, marked the beginning of an era of instant global communication. Telstar demonstrated the feasibility of using satellites to relay telephone and television signals across continents.

Satellite Internet: Today, satellites provide internet access to remote and underserved regions, bridging the digital divide and enabling access to education, healthcare, and economic opportunities.

The Role of Microwaves in Cellular Networks

Microwaves are the backbone of modern cellular networks, including 4G LTE and 5G technologies. These networks rely on microwave frequencies to transmit voice, data, and video over wireless connections.

4G and 5G Networks: The transition from 4G to 5G networks has brought significant improvements in speed, latency, and capacity. 5G technology uses millimeter-wave bands, a subset of microwaves, to achieve faster data transmission rates, supporting applications like autonomous vehicles, smart cities, and the Internet of Things (IoT).

Wi-Fi and Bluetooth: Wi-Fi and Bluetooth technologies also operate in the microwave spectrum, enabling wireless connectivity for personal devices like smartphones, laptops, and smart home systems.

Microwaves in Space Exploration

Microwaves are critical for space exploration, as they are used for communication between spacecraft and ground control. NASA's Deep Space Network (DSN) relies on microwave signals to send and receive data from distant space probes, such as the Mars rovers and interplanetary missions.

The Voyager Missions: The Voyager probes, launched in 1977, continue to communicate with Earth using microwave signals, even as they journey beyond our solar system. These communications provide invaluable data about the outer reaches of space.

The Role of Microwaves in Medicine and Industry

Medical Imaging and Treatment

Microwave technology has found applications in the medical field, particularly in imaging and treatment.

Microwave Ablation: This technique uses microwaves to generate heat and destroy cancerous tissue, offering a minimally invasive option for treating tumors.

Microwave Imaging: Emerging technologies in microwave imaging are being explored for breast cancer detection, providing a safer alternative to X-rays.

Industrial Applications: From Cooking to Material Testing

Microwaves are also used in various industrial processes, such as drying, material testing, and moisture analysis.

Microwave Ovens: The invention of the microwave oven revolutionized cooking, using microwaves to heat food quickly and efficiently.

Nondestructive Testing (NDT): Microwaves are used in NDT techniques to inspect materials and structures without causing damage, ensuring quality control in industries like aerospace and construction.

The Future of Radio and Microwave Technologies

As we move further into the 21st century, radio and microwave technologies continue to evolve, offering new possibilities for communication, exploration, and innovation.

Emerging Technologies

Terahertz Communication: Researchers are exploring terahertz frequencies, which lie between microwaves and infrared on the electromagnetic spectrum, for ultra-high-speed data transmission. This could lead to even faster wireless networks, supporting the growing demand for data-intensive applications like virtual reality and holographic communication.

Quantum Communication: The integration of microwaves with quantum technologies is opening new frontiers in secure communication. Quantum microwave links could provide encrypted communication channels that are resistant to hacking, ensuring data privacy in an increasingly digital world.

The Role of AI and Machine Learning

Artificial intelligence (AI) and machine learning are being integrated with radio and microwave technologies to optimize network performance, enhance spectrum management, and improve the efficiency of communication systems. This convergence of technologies is paving the way for intelligent networks that can adapt to changing conditions in real time.

The discovery and application of radio and microwave technologies have had a profound impact on modern society, enabling everything from global communication to space exploration. These invisible rays have connected the world, transformed industries, and opened new frontiers in science and technology. As we continue to harness the power of these rays, the future promises even more breakthroughs that will shape the way we live, work, and explore the universe.

By understanding the science behind radio and microwaves, we can appreciate their transformative role in communication and innovation. From the early days of wireless telegraphy to the high-speed 5G networks of today, these technologies have bridged the gaps between people, places, and ideas, truly revolutionizing the way we interact with the world around us.

18. ENVIRONMENTAL MONITORING

Using Rays to Study and Protect Our Planet

Environmental monitoring plays a crucial role in understanding the health of our planet. The need to protect Earth's delicate ecosystems, manage natural resources sustainably, and mitigate the effects of climate change has never been more urgent. This is where the invisible rays of the electromagnetic spectrum—ranging from gamma rays to radio waves—come into play. These rays are invaluable tools in monitoring various environmental parameters, from air and water quality to deforestation and climate change. In this chapter of "Harnessing the Invisible: The Science and Applications of Rays from Alpha to X," we explore how different types of rays are employed in environmental monitoring, helping scientists study and protect our planet.

The Role of Rays in Environmental Monitoring

The Science Behind Rays

Before diving into the specific applications, it is essential to understand the fundamental science behind rays. Rays are part of the electromagnetic spectrum, which includes a range of waves with different wavelengths and frequencies. The spectrum is divided into several categories, such as gamma rays, X-rays, ultraviolet (UV) rays, visible light, infrared (IR) rays, microwaves, and radio waves.

Each type of ray has unique properties that make it suitable for different environmental monitoring applications:

Gamma Rays: These have the highest energy and shortest wavelengths, allowing them to penetrate dense materials. They are useful for detecting radioactive contamination and soil analysis.

X-Rays: These are often used in soil and water quality assessments due to their ability to detect heavy metals and pollutants.

Ultraviolet Rays: UV rays are instrumental in studying the ozone layer and measuring solar radiation.

Infrared Rays: Infrared technology is widely used for monitoring vegetation health, water quality, and atmospheric conditions.

Microwaves: These are effective in weather forecasting, soil moisture analysis, and tracking deforestation.

Radio Waves: Radio waves are used in remote sensing technologies like radar and are essential for tracking changes in Earth's surface and atmosphere.

Why Environmental Monitoring Matters

Environmental monitoring is the systematic collection of data to observe and understand the natural environment. It is vital for several reasons:

Climate Change Analysis: Monitoring the environment helps scientists track the effects of global warming, such as rising temperatures, melting glaciers, and shifting weather patterns.

Pollution Control: By measuring air and water quality, we can identify pollution sources, assess their impact, and implement measures to reduce harmful emissions.

Biodiversity Conservation: Tracking changes in ecosystems and wildlife populations helps in conservation efforts and ensures the protection of endangered species.

Natural Disaster Management: Monitoring systems can provide early warning for natural disasters like hurricanes, earthquakes, and tsunamis, allowing for timely evacuations and disaster preparedness.

Applications of Different Rays in Environmental Monitoring

1. Gamma Rays: Detecting Radioactive Contamination

Gamma rays are the most energetic form of electromagnetic radiation and are capable of penetrating deep into the Earth's surface. This makes them particularly useful for detecting radioactive contamination in the environment.

Applications:

Radiation Monitoring: Gamma-ray spectrometers are used to monitor radiation levels in the environment, particularly around nuclear power plants, medical facilities, and areas affected by nuclear accidents like Chernobyl and Fukushima.

Soil Analysis: Gamma rays can be used to analyze soil composition, detecting the presence of radioactive isotopes, heavy metals, and other contaminants that may pose a risk to agriculture and human health.

Cosmic Ray Detection: Gamma-ray detectors are also used in astrophysics to study cosmic rays, which can provide insights into solar activity and its impact on Earth's atmosphere.

2. X-Rays: Assessing Soil and Water Quality

X-ray fluorescence (XRF) technology is a powerful tool for assessing the quality of soil and water. By measuring the characteristic X-rays emitted by materials, scientists can identify the presence of heavy metals and other pollutants.

Applications:

Water Quality Monitoring: X-rays can detect harmful contaminants like lead, mercury, and arsenic in water supplies, ensuring safe drinking water for communities.

Soil Contamination Studies: XRF analysis helps in identifying soil pollution caused by industrial waste, mining activities, and agricultural runoff. This is crucial for maintaining soil health and ensuring sustainable agricultural practices.

3. Ultraviolet Rays: Studying the Ozone Layer and Solar Radiation

Ultraviolet (UV) rays play a significant role in studying atmospheric conditions, particularly the ozone layer, which protects Earth from harmful solar radiation.

Applications:

Ozone Layer Monitoring: Satellites equipped with UV sensors measure the concentration of ozone in the stratosphere. This data is crucial for tracking ozone depletion, especially over the polar regions, and understanding the impact of human activities on ozone recovery.

Solar Radiation Measurement: UV radiation measurements help scientists assess the intensity of solar radiation reaching Earth's surface. This information is vital for studying the effects of solar radiation on ecosystems, human health, and climate change.

4. Infrared Rays: Monitoring Vegetation, Water Quality, and Climate

Infrared (IR) rays are widely used in remote sensing to monitor environmental conditions. The ability of infrared radiation to detect heat makes it ideal for various ecological studies.

Applications:

Vegetation Health: Infrared sensors on satellites and drones are used to monitor plant health by measuring the Normalized Difference Vegetation Index (NDVI). Healthy vegetation reflects more infrared light, while stressed or damaged vegetation reflects less.

Water Quality: Infrared technology is also employed in assessing water quality by detecting changes in temperature, salinity, and the presence of pollutants in rivers, lakes, and oceans.

Climate Monitoring: Infrared imaging helps track atmospheric temperatures, cloud cover, and greenhouse gas concentrations, providing essential data for climate change research.

5. Microwaves: Weather Forecasting and Natural Resource Management

Microwaves penetrate clouds, fog, and vegetation, making them useful for all-weather environmental monitoring.

Applications:

Weather Forecasting: Microwave radiometers on satellites measure atmospheric moisture, temperature profiles, and precipitation levels, improving the accuracy of weather forecasts.

Soil Moisture Analysis: Microwave sensors help in assessing soil moisture content, which is crucial for agricultural planning, drought management, and understanding the hydrological cycle.

Deforestation Tracking: Radar systems using microwaves are effective in mapping forest cover and detecting changes due to deforestation, logging, and forest fires.

6. Radio Waves: Remote Sensing and Environmental Exploration

Radio waves, particularly in the form of radar technology, have numerous applications in environmental monitoring. Radar systems can operate in various weather conditions and provide detailed information about Earth's surface.

Applications:

Earth Observation: Synthetic Aperture Radar (SAR) is used to create high-resolution images of Earth's surface, aiding in land use planning, urban development, and environmental conservation.

Ocean Monitoring: Radio waves are used to study ocean currents, sea surface temperatures, and wave heights, which are essential for marine navigation, fisheries management, and climate research.

Wildlife Tracking: Radio telemetry, using radio waves, is employed to track animal movements, helping scientists study migration patterns, habitat use, and conservation efforts.

Case Studies: How Rays are Transforming Environmental Monitoring

Case Study 1: Monitoring the Amazon Rainforest

The Amazon rainforest, often referred to as the "lungs of the Earth," plays a vital role in regulating the global climate. However, it is under threat from deforestation, illegal logging, and wildfires. Satellites equipped with infrared and microwave sensors are used to monitor the health of the

Amazon, detecting areas of deforestation, forest degradation, and illegal activities. This data is critical for conservation efforts and for enforcing environmental regulations.

Case Study 2: Tracking Air Quality in Urban Areas

Urban air pollution is a major public health concern, contributing to respiratory diseases and premature deaths. Remote sensing technology, using ultraviolet and infrared rays, is employed to measure air quality in real time. Satellites like the Sentinel-5P use these rays to detect pollutants such as nitrogen dioxide, sulfur dioxide, and particulate matter, enabling cities to implement strategies for reducing air pollution.

Case Study 3: Studying Climate Change in the Arctic

The Arctic region is warming at twice the rate of the rest of the planet, leading to melting ice caps and rising sea levels. Infrared and microwave satellites are used to monitor the extent of sea ice, glacier retreat, and temperature changes in the Arctic. This information is crucial for understanding the impact of climate change on polar ecosystems and for predicting future sea level rise.

The Future of Environmental Monitoring with Rays

Advancements in technology continue to expand the capabilities of environmental monitoring. The integration of artificial intelligence (AI) with remote sensing data allows for more accurate predictions and real-time analysis. Additionally, the development of nanosatellites and CubeSats, equipped with sensors across the electromagnetic spectrum, is making environmental monitoring more accessible and cost-effective.

Emerging Trends:

Hyperspectral Imaging: This technology captures data across hundreds of narrow spectral bands, providing detailed information about land, water, and vegetation. Hyperspectral imaging is being explored for applications like monitoring crop health, detecting oil spills, and studying coral reefs.

Quantum Sensing: Quantum technologies are being developed to enhance the sensitivity of environmental monitoring instruments, allowing for the detection of minute changes in temperature, magnetic fields, and chemical compositions.

Citizen Science: Increasingly, communities and citizen scientists are contributing to environmental monitoring efforts using low-cost sensors and mobile apps, creating a more decentralized and inclusive approach to data collection.

The invisible rays of the electromagnetic spectrum are powerful tools for studying and protecting our planet. From detecting radioactive contamination to tracking deforestation and monitoring air quality, these rays provide critical data that helps scientists, policymakers, and communities make informed decisions for a sustainable future. As technology continues to advance, the role of rays in environmental monitoring will only become more significant, enabling us to address the challenges of climate change, pollution, and biodiversity loss with greater precision and effectiveness.

By harnessing the invisible, we can ensure a healthier, more resilient planet for generations to come.

19. SECURITY AND DEFENSE

How Rays Keep Us Safe from Hidden Threats

In an increasingly complex and interconnected world, the need for robust security and defense measures is more critical than ever. From protecting borders and critical infrastructure to ensuring public safety, modern security solutions have evolved to counter an array of hidden threats. One of the most powerful tools in this arsenal is the use of various rays from the electromagnetic spectrum, which have proven to be invaluable in detecting, identifying, and neutralizing threats that are otherwise invisible to the naked eye.

In this chapter of "Harnessing the Invisible: The Science and Applications of Rays from Alpha to X," we delve into how different types of rays, including X-rays, gamma rays, infrared rays, and microwaves, are harnessed to enhance security and defense capabilities. These technologies are not only essential in countering traditional threats such as weapons smuggling and border breaches but also play a vital role in the detection of chemical, biological, radiological, and nuclear (CBRN) hazards, thereby keeping our societies safer.

The Electromagnetic Spectrum in Security and Defense

The electromagnetic spectrum encompasses a wide range of rays, each with unique properties that make them suitable for various security applications:

X-rays: Known for their ability to penetrate solid objects, X-rays are widely used in security scanning and baggage inspection.

Gamma rays: With their high energy, gamma rays are employed in radiological detection and materials analysis, particularly in identifying radioactive substances.

Infrared rays: These rays detect heat signatures, making them ideal for night vision, surveillance, and thermal imaging.

Microwaves: Microwaves are crucial in radar systems for monitoring airspace, maritime environments, and ground activities.

These rays provide diverse solutions to security challenges, from detecting concealed weapons to scanning cargo containers for illicit materials.

X-Rays: Unveiling the Concealed

X-Ray Security Scanning

X-ray technology is one of the most widely used methods in security applications, particularly in airports, seaports, and border checkpoints. The ability of X-rays to penetrate various materials makes them ideal for scanning luggage, packages, and even vehicles to detect hidden contraband, explosives, and weapons.

Applications:

Baggage Screening: Airport security relies heavily on X-ray scanners to inspect passengers' luggage for prohibited items. These scanners can differentiate between organic and inorganic materials, making it easier to spot potential threats like explosives or firearms.

Cargo and Container Inspection: Large-scale X-ray systems are employed at ports of entry to inspect shipping containers without the need to unload them. This not only saves time but also ensures the detection of illicit goods, such as drugs, counterfeit products, and smuggled weapons.

Vehicle Scanning: Border security often uses mobile X-ray scanners to inspect vehicles for hidden compartments that may contain contraband or illegal immigrants.

Advanced X-Ray Technologies

Recent advancements in X-ray technology have led to the development of dual-energy and backscatter X-ray systems, which provide more detailed imaging. Dual-energy X-rays use two different energy levels to distinguish between different types of materials, while backscatter X-rays capture scattered radiation, revealing objects on or near the body, such as concealed knives or explosive belts.

Case Study: X-Ray Systems in Airport Security

Following the 9/11 attacks, airports around the world enhanced their security measures by incorporating advanced X-ray technologies. These systems are capable of detecting non-metallic threats, such as plastic explosives, which traditional metal detectors might miss. The implementation of full-body scanners using millimeter-wave and backscatter X-rays has significantly improved the ability to detect hidden threats on passengers.

Gamma Rays: Detecting Radioactive Threats

Gamma-Ray Spectroscopy

Gamma rays, with their high-energy and deep penetration capabilities, are particularly useful in the detection of radioactive substances. Gamma-ray spectroscopy is a technique used to identify the presence of specific isotopes based on their unique energy signatures.

Applications:

Nuclear Security: Gamma-ray detectors are deployed at critical infrastructure sites, such as nuclear power plants, to detect and prevent unauthorized access to radioactive materials. They are also used at ports and border crossings to scan cargo for illicit radioactive materials that could be used in dirty bombs or other radiological weapons.

Radiation Detection: Law enforcement agencies use portable gamma-ray detectors to identify radioactive sources in public spaces, ensuring quick response to potential radiological threats.

Homeland Security and Counterterrorism

Gamma-ray imaging plays a crucial role in homeland security by providing the means to detect radiological dispersal devices (RDDs) and improvised nuclear devices (INDs). Early detection of these threats is essential in preventing terrorist attacks and mitigating the impact of radiological exposure.

Case Study: Gamma-Ray Detection in CBRN Defense

In the aftermath of incidents like the poisoning of Alexander Litvinenko with polonium-210, governments have ramped up efforts to detect radioactive materials in public areas. Gamma-ray spectrometers are now standard equipment for emergency response teams dealing with CBRN threats, providing a non-invasive way to identify and neutralize radiological hazards.

Infrared Rays: Enhancing Surveillance and Detection

Infrared Surveillance and Night Vision

Infrared (IR) technology detects heat signatures emitted by objects, making it an invaluable tool for surveillance, particularly in low-light or nighttime conditions. IR cameras can identify individuals, vehicles, and other heat-emitting objects, even in complete darkness.

Applications:

Border Patrol: Infrared cameras are used by border security agencies to monitor illegal crossings and detect human activity in remote areas. These systems are effective in spotting movement over vast terrains, including deserts and forests, where traditional surveillance might fail.

Search and Rescue: Infrared imaging helps locate missing persons, especially in challenging environments like dense forests or disaster-stricken areas. The technology can detect body heat, enabling rescue teams to find survivors quickly.

Building Security: Infrared sensors are widely used in motion detectors and surveillance systems to secure critical infrastructure and high-value assets.

Thermal Imaging in Law Enforcement

Thermal imaging cameras have become standard tools for law enforcement agencies, providing situational awareness during operations. These cameras can detect the heat signatures of individuals hiding behind walls or foliage, aiding in hostage rescues and tactical operations.

Case Study: Infrared Technology in Military Operations

The military extensively uses infrared technology for night vision goggles and thermal imaging scopes, giving soldiers an advantage in low-visibility conditions. During the Gulf War, infrared-guided missile systems were pivotal in targeting and neutralizing enemy equipment, demonstrating the strategic importance of IR technology in modern warfare.

Microwaves: The Power of Radar in Security

Radar Systems for Air and Maritime Security

Microwaves are the backbone of radar systems, which are critical for monitoring airspace and maritime environments. Radar technology uses the reflection of microwaves to detect the position, speed, and trajectory of objects, making it essential for security and defense.

Applications:

Air Traffic Control: Radars are used to track aircraft movements, ensuring safe navigation in crowded airspace. They also play a vital role in detecting unauthorized flights and potential aerial threats.

Maritime Surveillance: Coastal security agencies use radar to monitor shipping lanes, detect unauthorized vessels, and prevent smuggling, piracy, and illegal fishing activities.

Ground Surveillance: Radar systems are also deployed for perimeter security around military bases and critical infrastructure to detect and respond to intrusions.

Microwave Technology in Counter-Drone Measures

With the rise of drone technology, there is an increasing need for counter-drone measures to protect sensitive areas from aerial threats. Microwave systems are being developed to detect, track, and disable unauthorized drones, preventing them from being used for surveillance or attacks.

Case Study: Radar Systems in Anti-Drone Defense

The use of drones by terrorists and hostile states has led to the development of microwave-based anti-drone systems. These systems can jam the communication signals of drones, forcing them to land or return to their point of origin, thus preventing potential threats to critical infrastructure and public events.

Future Trends in Security and Defense Using Rays

Advances in Terahertz Technology

Terahertz (THz) waves, situated between microwaves and infrared rays in the electromagnetic spectrum, are emerging as a powerful tool in security. Terahertz imaging can penetrate materials like clothing and paper, making it useful for detecting hidden weapons and contraband without physical contact.

Potential Applications:

Non-Invasive Security Screening: Terahertz scanners can provide detailed images of concealed items without exposing individuals to harmful radiation, offering a safer alternative to X-ray scanners.

Chemical Detection: Terahertz spectroscopy can identify chemical signatures, which is crucial for detecting explosives, narcotics, and hazardous substances in real-time.

Quantum Sensing for Enhanced Detection

Quantum technology is set to revolutionize security and defense applications. Quantum sensors, which exploit the principles of quantum mechanics, offer unprecedented sensitivity in detecting minute changes in electromagnetic fields, temperature, and gravitational forces.

Emerging Capabilities:

Stealth Detection: Quantum radar systems are being developed to detect stealth aircraft and submarines, which are designed to evade conventional radar systems.

Seismic Monitoring: Quantum sensors can detect underground activities, such as tunneling, which are often used by smugglers and infiltrators.

The invisible rays of the electromagnetic spectrum are not just tools of scientific curiosity but are actively transforming the way we ensure security and defense. From X-ray scanners that keep our airports safe to infrared cameras that enable night-time surveillance, these rays are essential in detecting and neutralizing hidden threats. As technology continues to advance, new applications of rays in security and defense will undoubtedly emerge, providing even more sophisticated methods to protect our societies.

The future of security lies in our ability to harness the invisible, making the world a safer place by detecting the unseen dangers that threaten our way of life. Whether it's through the development of quantum sensors or the deployment of terahertz imaging, the use of rays in security and defense will continue to be a cornerstone of modern safety protocols, ensuring that we can face the challenges of an uncertain world with confidence and resilience.

20. THE FUTURE OF RAY TECHNOLOGY

Innovations on the Horizon

In the dynamic landscape of modern science and technology, rays from the electromagnetic spectrum have played a pivotal role in transforming our world. From radio waves that revolutionized communication to X-rays that unveiled the mysteries of the human body, the applications of these invisible rays are vast and diverse. As we stand on the brink of a new era in scientific discovery, the future of ray technology promises even more groundbreaking innovations. These advancements are set to redefine various fields, including medicine, security, communication, industry, and space exploration.

In this chapter of "Harnessing the Invisible: The Science and Applications of Rays from Alpha to X," we explore the future of ray technology, delving into emerging innovations that are set to push the boundaries of what is possible. We will examine how advancements in ray-based technologies are poised to reshape our understanding of the world, enhance our capabilities, and address some of the most pressing challenges facing humanity.

The Evolution of Ray Technology: A Brief Overview

The journey of ray technology began with the discovery of radio waves by Heinrich Hertz in the late 19th century, followed by Wilhelm Röntgen's discovery of X-rays in 1895. These groundbreaking discoveries opened the doors to new possibilities in communication, medical diagnostics, and industrial applications. Over the years, scientists have harnessed different types of rays, such as gamma rays, infrared rays, microwaves, and ultraviolet rays, each with its unique properties and applications.

The continuous advancement of ray technology has been driven by the development of sophisticated devices and techniques, including CT scanners, MRIs, infrared cameras, and radio telescopes. These technologies have not only expanded our scientific knowledge but have also led to significant societal benefits. As we look to the future, the next generation of ray technologies promises to be even more transformative, unlocking new applications and capabilities that were once thought to be science fiction.

The Next Frontier in Medical Applications

1. Advanced Imaging Techniques: The Era of Hyper-Resolution

Medical imaging has long been one of the most impactful applications of ray technology. However, the future holds the promise of even more advanced imaging techniques that will provide unprecedented levels of detail and accuracy.

Innovations on the Horizon:

Quantum-Enhanced Imaging: Leveraging the principles of quantum mechanics, researchers are developing imaging techniques that use entangled photons to achieve higher resolution and sensitivity. This could revolutionize the early detection of diseases, such as cancer, by providing clearer images of tumors and other abnormalities at a molecular level.

Terahertz Imaging: Terahertz (THz) waves, situated between infrared and microwaves in the electromagnetic spectrum, are non-ionizing and can penetrate various materials without causing harm. Terahertz imaging is poised to become a powerful tool in medical diagnostics, capable of detecting skin cancers, dental issues, and even monitoring burn wounds without physical contact.

Photoacoustic Imaging: Combining the advantages of optical imaging and ultrasound, photoacoustic imaging uses laser-induced ultrasound waves to create high-resolution images of tissues. This technique is expected to play a crucial role in visualizing blood vessels, detecting tumors, and monitoring oxygen levels in real-time.

2. Targeted Cancer Treatment: The Rise of Ray-Based Therapies

Radiation therapy has been a cornerstone in cancer treatment for decades. However, advancements in ray technology are leading to more precise and less invasive treatment options.

Innovations on the Horizon:

Proton Beam Therapy: Unlike traditional radiation therapy, which uses X-rays, proton therapy uses positively charged particles (protons) that deliver targeted radiation to tumors with minimal damage to surrounding healthy tissues. Future developments in proton therapy aim to increase its accessibility and effectiveness, making it a viable option for more types of cancers.

Gamma Knife and CyberKnife: These technologies use focused beams of gamma rays to treat brain tumors and other abnormalities with extreme precision, often eliminating the need for invasive surgery. The next generation of these devices promises to be even more accurate, reducing treatment times and improving patient outcomes.

Nanoparticle-Assisted Radiotherapy: Researchers are exploring the use of nanoparticles to enhance the effectiveness of radiation therapy. These nanoparticles can be engineered to target cancer cells specifically, increasing the dose of radiation delivered to the tumor while sparing healthy tissues.

Enhancing Security and Defense with Ray Technology

3. Next-Generation Scanning and Detection Systems

In the realm of security and defense, the ability to detect threats quickly and accurately is crucial. Ray technology continues to evolve, providing more sophisticated solutions for scanning and detection.

Innovations on the Horizon:

Terahertz Security Scanners: Unlike X-rays, terahertz waves are non-ionizing and safe for human exposure. Future terahertz scanners will be capable of detecting concealed weapons, explosives, and illicit substances with higher sensitivity and speed, making them ideal for airport security, border control, and public safety.

Gamma-Ray Spectroscopy for Nuclear Security: As the threat of nuclear proliferation increases, there is a growing need for advanced detection systems that can identify radioactive materials. Enhanced gamma-ray spectroscopy devices are being developed to detect and identify specific isotopes, even in shielded containers, providing an essential tool for nuclear security.

Directed Energy Weapons: Military applications of ray technology include the development of directed energy weapons (DEWs), such as high-energy lasers and microwave weapons. These systems can be used for disabling enemy drones, vehicles, and electronic equipment, offering a non-lethal means of neutralizing threats.

4. Infrared and Thermal Imaging for Surveillance

The use of infrared rays in surveillance and reconnaissance is set to become even more sophisticated with the integration of artificial intelligence and machine learning.

Innovations on the Horizon:

AI-Powered Thermal Cameras: Future thermal imaging systems will be equipped with AI algorithms capable of analyzing heat signatures and identifying suspicious activities in real-time. This technology will enhance security in critical infrastructure, military bases, and urban areas.

Wearable Infrared Devices: Portable infrared scanners integrated into wearable devices could provide soldiers and first responders with real-time situational awareness, allowing them to detect hidden threats and navigate challenging environments more effectively.

Revolutionizing Communication and Connectivity

5. The Promise of 6G and Beyond

The evolution of wireless communication continues with the development of 6G technology, which will rely on higher frequency bands, including terahertz waves, to deliver ultra-fast data transmission.

Innovations on the Horizon:

Terahertz Communication Networks: Terahertz waves offer significantly higher bandwidth than existing communication systems, enabling data transfer rates that are orders of magnitude faster than 5G. This will pave the way for advanced applications such as holographic communication, real-time virtual reality, and ultra-low-latency Internet of Things (IoT) networks.

Satellite-Based Quantum Communication: The future of secure communication lies in the use of quantum rays to establish unhackable communication channels. Satellite-based quantum communication networks are being developed to provide global coverage, ensuring secure data transmission across long distances.

6. Enhancing Global Navigation with Ray Technology

Ray technology is also transforming global navigation systems, with advancements in radar and satellite-based positioning.

Innovations on the Horizon:

Quantum Radar: Using entangled photons, quantum radar systems promise to detect stealth aircraft and other objects that are invisible to conventional radar. This could enhance air traffic control, military surveillance, and search-and-rescue missions.

Pulsar-Based Navigation: Researchers are exploring the use of X-rays emitted by pulsars (neutron stars) for deep-space navigation. This method could provide more accurate positioning for spacecraft, enabling long-duration missions beyond our solar system.

Industrial and Environmental Applications

7. Nondestructive Testing and Material Analysis

The use of rays for nondestructive testing (NDT) is expanding, with innovations that improve the accuracy and efficiency of material analysis.

Innovations on the Horizon:

Advanced X-Ray Diffraction (XRD): Future XRD systems will provide faster and more detailed analysis of materials, helping industries optimize manufacturing processes and ensure product quality.

Neutron Radiography: Unlike X-rays, neutrons can penetrate heavy metals, making them ideal for inspecting critical components in aerospace, automotive, and energy sectors. Enhanced neutron imaging techniques will enable the detection of micro-defects and structural anomalies that are invisible to other methods.

8. Environmental Monitoring and Climate Science

Ray technology is playing an increasing role in environmental monitoring, helping scientists understand and address climate change.

Innovations on the Horizon:

Lidar for Atmospheric Studies: Light Detection and Ranging (Lidar) technology uses laser rays to study atmospheric conditions, such as cloud formation, pollution levels, and greenhouse gas concentrations. Advanced Lidar systems will provide more accurate data for climate models, enabling better predictions of weather patterns and environmental changes.

Gamma-Ray Soil Analysis: Portable gamma-ray spectrometers are being developed to assess soil composition, helping farmers optimize crop yields and reduce environmental impact.

Space Exploration: Rays as a Tool for Discovery

9. Exploring the Cosmos with High-Energy Rays

Ray technology is a cornerstone of modern astronomy, providing insights into the universe's most energetic phenomena.

Innovations on the Horizon:

Next-Generation X-Ray Telescopes: Future X-ray telescopes, equipped with adaptive optics and high-resolution sensors, will enable astronomers to study black holes, neutron stars, and supernovae with unprecedented detail.

Gamma-Ray Burst Detectors: Researchers are developing advanced gamma-ray detectors to study gamma-ray bursts (GRBs), which are among the most powerful explosions in the universe. Understanding GRBs could reveal new insights into the formation of galaxies and the behavior of matter under extreme conditions.

The Boundless Potential of Ray Technology

The future of ray technology is filled with possibilities that will redefine how we perceive, interact with, and protect our world. As we continue to push the boundaries of what is possible, the innovations on the horizon promise to bring about a new era of scientific discovery, technological advancement, and societal progress. From revolutionizing healthcare and communication to enhancing security and environmental protection, the potential applications of ray technology are limited only by our imagination.

The journey of harnessing the invisible rays is far from over. As we venture into the unknown, the next generation of ray-based technologies will illuminate new frontiers, enabling us to unlock the secrets of the universe and build a brighter, safer, and more connected world.

21. ETHICAL CONSIDERATIONS AND SAFETY REGULATIONS

Balancing Progress with Responsibility

The exploration and utilization of rays from the electromagnetic spectrum have brought about significant advancements in science, medicine, industry, and technology. From the groundbreaking discovery of X-rays to the development of sophisticated imaging techniques and the use of radiation in cancer therapy, the power of invisible rays has transformed our understanding of the world and improved our quality of life. However, with these advancements come complex ethical considerations and the need for robust safety regulations. As we continue to harness the potential of rays from alpha to X and beyond, it is crucial to balance progress with responsibility, ensuring that the benefits of these technologies do not come at the expense of human health, privacy, or the environment.

In this chapter of "Harnessing the Invisible: The Science and Applications of Rays from Alpha to X," we delve into the ethical considerations and safety regulations that guide the responsible use of ray technology. We explore the challenges and dilemmas faced by scientists, policymakers, and industries as they navigate the fine line between innovation and safety. We will also examine the frameworks and guidelines that have been established to protect individuals and society, ensuring that the pursuit of progress does not compromise our ethical values.

The Ethical Landscape of Ray Technology

1. The Dual-Use Dilemma: Scientific Advancements and Potential Misuse

One of the most pressing ethical concerns surrounding ray technology is its dual-use nature. Many applications of rays, such as X-rays, gamma rays, and microwaves, have both beneficial and harmful potential. For example, while ionizing radiation is a powerful tool in medical diagnostics and cancer treatment, it can also be used in harmful ways, such as in the development of nuclear weapons or covert surveillance technologies.

Ethical Considerations:

Medical vs. Military Use: The same technologies that enable life-saving cancer treatments can be weaponized for military purposes. This raises ethical questions about the development and regulation of technologies that can be used for both healing and harm.

Surveillance and Privacy: Technologies like millimeter-wave scanners and infrared cameras, which are used for security screening and surveillance, pose significant privacy concerns. Balancing the need for security with the right to privacy is a critical ethical challenge.

Scientific Freedom vs. Social Responsibility: Researchers and scientists often face the dilemma of advancing knowledge while being mindful of the potential societal impacts of their work. Ethical guidelines are needed to ensure that scientific discoveries are used for the greater good, rather than causing harm.

2. Radiation in Medicine: The Ethics of Risk vs. Benefit

The use of radiation in medical applications has undoubtedly saved countless lives, particularly in the diagnosis and treatment of diseases like cancer. However, the exposure to ionizing radiation, even in controlled medical settings, carries inherent risks, such as the potential to cause secondary cancers or genetic mutations.

Ethical Considerations:

Informed Consent: Patients undergoing diagnostic imaging or radiation therapy must be fully informed of the risks and benefits associated with the procedure. Informed consent is an ethical cornerstone in medical practice, ensuring that patients have the autonomy to make decisions about their health.

Balancing Harm and Benefit: Medical practitioners must weigh the potential benefits of radiation-based treatments against the risks of radiation exposure. This ethical principle, known as "primum non nocere" (first, do no harm), is central to patient care.

Access and Equity: While advanced radiation therapies, such as proton beam therapy, offer improved outcomes for cancer patients, they are often expensive and not widely accessible. Ethical considerations around healthcare equity and access are crucial in ensuring that life-saving treatments are available to all, regardless of socioeconomic status.

3. Environmental Impact: The Ethics of Radiation Pollution

The industrial use of rays, such as gamma rays for sterilization or X-rays for material inspection, has raised concerns about environmental pollution and its impact on human and ecological health. The disposal of radioactive materials, particularly from nuclear power plants, poses long-term risks to the environment.

Ethical Considerations:

Environmental Stewardship: Industries using radioactive materials have an ethical obligation to minimize their environmental footprint and ensure the safe disposal of waste. This includes investing in technologies that reduce radiation pollution and adopting best practices for waste management.

Intergenerational Responsibility: The long half-life of certain radioactive isotopes means that the impact of radiation pollution can last for thousands of years. Ethical considerations around intergenerational justice require that current generations take responsibility for protecting future generations from the harmful effects of radiation.

Precautionary Principle: In cases where the environmental impact of radiation exposure is uncertain, the precautionary principle should guide decision-making. This principle emphasizes the need to err on the side of caution, particularly when there is a lack of scientific consensus on the potential risks.

Safety Regulations and Guidelines: Protecting Health and Well-being

4. International Frameworks for Radiation Protection

To address the ethical concerns associated with ray technology, various international organizations have established guidelines and frameworks to ensure the safe use of radiation. The International Commission on Radiological Protection (ICRP) and the International Atomic Energy Agency (IAEA) are two key organizations that play a crucial role in setting safety standards.

Key Regulatory Principles:

Justification: Any application of radiation must be justified, meaning that the benefits of its use must outweigh the risks. This principle is particularly important in medical settings, where unnecessary exposure to radiation should be avoided.

Optimization (ALARA Principle): The principle of "As Low As Reasonably Achievable" (ALARA) emphasizes the need to minimize radiation exposure by optimizing procedures and technologies. This applies to both medical and industrial applications.

Dose Limits: Regulatory bodies have established dose limits for occupational exposure, public exposure, and patient exposure. These limits are designed to protect individuals from the harmful effects of ionizing radiation while allowing the benefits of its use.

5. National Regulations and Compliance

In addition to international guidelines, individual countries have their own regulations governing the use of radiation. These regulations vary depending on the specific application, such as medical imaging, nuclear power, or industrial testing.

Examples of National Regulatory Agencies:

United States: The Nuclear Regulatory Commission (NRC) and the Environmental Protection Agency (EPA) are responsible for regulating radiation safety in the United States.

European Union: The European Atomic Energy Community (Euratom) sets radiation protection standards for member states.

Japan: The Nuclear Regulation Authority (NRA) oversees radiation safety, particularly in the wake of the Fukushima nuclear disaster.

6. Technological Innovations for Enhanced Safety

Advancements in technology are playing a significant role in enhancing radiation safety. Innovations such as real-time dosimetry, automated monitoring systems, and AI-driven safety protocols are helping to reduce radiation exposure and improve compliance with safety regulations.

Examples of Safety Innovations:

Digital Dosimeters: These devices provide real-time monitoring of radiation exposure, allowing healthcare workers and industrial personnel to track their cumulative dose and take preventive measures.

AI-Powered Imaging Systems: Artificial intelligence is being used to optimize radiation doses in medical imaging, ensuring that patients receive the lowest possible dose while still obtaining high-quality images.

Remote Radiation Monitoring: In industrial and environmental settings, remote sensors and drones are being used to monitor radiation levels in real-time, reducing the need for human exposure in hazardous environments.

Balancing Innovation and Responsibility: The Path Forward

7. Ethical Frameworks for Emerging Ray Technologies

As new applications of ray technology continue to emerge, there is a growing need for ethical frameworks that guide their development and use. This includes the ethical implications of using rays in areas such as genetic engineering, space exploration, and quantum communication.

Future Ethical Considerations:

Genetic Editing with Rays: The use of radiation to induce genetic mutations raises ethical questions about the potential for unintended consequences, particularly in the field of human genetics.

Space Exploration: The use of cosmic rays for space propulsion and deep-space communication presents ethical challenges related to the impact on astronauts' health and the potential for contamination of other planets.

Quantum Ray Technologies: As quantum rays are harnessed for secure communication and computing, ethical considerations around data privacy and cybersecurity will become increasingly important.

8. Public Engagement and Ethical Governance

The ethical use of ray technology requires not only the input of scientists and policymakers but also the involvement of the public. Transparent communication, public engagement, and ethical governance are essential for building trust and ensuring that the benefits of ray technology are shared equitably.

Strategies for Ethical Governance:

Public Education: Increasing public awareness of the benefits and risks of ray technology can help build trust and support for its responsible use. This includes clear communication about the safety measures in place and the potential societal impacts of new technologies.

Ethical Review Boards: The establishment of ethical review boards can provide oversight for research and development projects involving ray technology. These boards can evaluate the ethical implications of proposed technologies and ensure that they align with societal values.

Stakeholder Collaboration: Collaboration between scientists, industry leaders, policymakers, and community organizations is essential for developing ethical guidelines that balance innovation with responsibility.

Conclusion: Navigating the Ethical Landscape of Ray Technology

The future of ray technology holds immense promise for improving human health, enhancing security, and expanding our understanding of the universe. However, as we continue to explore the potential of rays from alpha to X, it is imperative that we navigate the ethical landscape with care and responsibility. By adhering to ethical principles, implementing robust safety regulations, and fostering public trust, we can ensure that the benefits of ray technology are realized while minimizing the risks to individuals, society, and the environment.

As we move forward, the challenge will be to strike a balance between progress and responsibility, ensuring that the pursuit of scientific discovery and technological advancement does not compromise our ethical values. By doing so, we can harness the invisible rays of the electromagnetic spectrum to build a brighter, safer, and more sustainable future for all.